Your dog and you ...

Understanding the canine psyche

Hubble & Hattie

Gill Garratt

www.hubbleandhattie.com

The Hubble & Hattie imprint was launched in 2009 and is named in memory of two very special Westie sisters owned by Veloce's proprietors.
Since the first book, many more have been added to the list, all with the same underlying objective: to be of real benefit to the species they cover, at the same time promoting compassion, understanding and respect between all animals (including human ones!)
All Hubble & Hattie publications offer ethical, high quality content and presentation, plus great value for money.

More titles from Hubble and Hattie

Among the Wolves: Memoirs of a wolf handler (Shelbourne)

Animal Grief: How animals mourn (Alderton)

Because this is our home ... the story of a cat's progress (Bowes)

Camper vans, ex-pats & Spanish Hounds: from road trip torescue – the strays of Spain (Coates & Morris)

Cat Speak: recognising & understanding behaviour (Rauth-Widmann)

Clever dog! Life lessons from the world's most successful animal (O'Meara)

Complete Dog Massage Manual, The – Gentle Dog Care (Robertson)

Dieting with my dog: one busy life, two full figures ... andm unconditional love (Frezon)

Dinner with Rover: delicious, nutritious meals for you and your dog to share (Paton-Ayre)

Dog Cookies: healthy, allergen-free treat recipes for your dog (Schöps)

Dog-friendly Gardening: creating a safe haven for you and your dog (Bush)

Dog Games – stimulating play to entertain your dog and you (Blenski)

Dog Relax – relaxed dogs, relaxed owners (Pilguj)

Dog Speak: recognising & understanding behaviour (Blenski)

Dogs on Wheels: travelling with your canine companion (Mort)

Emergency First Aid for dogs: at home and away (Bucksch)

Exercising your puppy: a gentle & natural approach – Gentle Dog Care (Robertson & Pope)

Fun and Games for Cats (Seidl)

Gymnastricks: Targeted muscle training for dogs (Mayer)

Helping minds meet – skills for a better life with your dog (Zulch & Mills)

Know Your Dog – The guide to a beautiful relationship (Birmelin)

Life skills for puppies – laying the foundation of a loving, lasting relationship (Zuch & Mills)

Living with an Older Dog – Gentle Dog Care (Alderton & Hall)

Miaow! Cats really are nicer than people! (Moore)

My cat has arthritis – but lives life to the full! (Carrick)

My dog has arthritis – but lives life to the full! (Carrick)

My dog has cruciate ligament injury – but lives life to the full! (Haüsler & Friedrich)

My dog has epilepsy – but lives life to the full! (Carrick)

My dog has hip dysplasia – but lives life to the full! (Haüsler & Friedrich)

My dog is blind – but lives life to the full! (Horsky)

My dog is deaf – but lives life to the full! (Willms)

My Dog, my Friend: heart-warming tales of canine companionship from celebrities and other extraordinary people (Gordon)

No walks? No worries! Maintaining wellbeing in dogs on restricted exercise (Ryan & Zulch)

Partners – Everyday working dogs being heroes every day (Walton)

Smellorama – nose games for dogs (Theby)

Swim to recovery: canine hydrotherapy healing – Gentle Dog Care (Wong)

The Truth about Wolves and Dogs: dispelling the myths of dog training (Shelbourne)

Waggy Tails & Wheelchairs (Epp)

Walking the dog: motorway walks for drivers & dogs (Rees)

Winston ... the dog who changed my life (Klute)

You and Your Border Terrier – The Essential Guide (Alderton)

You and Your Cockapoo – The Essential Guide (Alderton)

Your dog and you – understanding the canine psyche (Garratt)

For post publication news, updates and amendments relating to this book please visit www.hubbleandhattie.com/extras/HH4738

Disclaimer

Please note that no dog was deliberately frightened during photographic sessions. The images used to depict dogs feeling worried about specific situations were taken whilst the animals were exploring the novel environment of the studio setup. These images were later modified to enable us to illustrate the points we wished to make. Please also note that all body language is context-specific, and individual signals can mean different things in different contexts.

First published in April 2015 by Veloce Publishing Limited, Veloce House, Parkway Farm Business Park, Middle Farm Way, Poundbury, Dorchester, Dorset, DT1 3AR, England.
Fax 01305 250479/e-mail info@hubbleandhattie.com/web www.hubbleandhattie.com. ISBN: 978-1-845847-38-8 UPC: 6-36847-04738-2. © Gillian Garrett, Tom Walters and Veloce Publishing Ltd 2015

Acknowledgements

I would like to thank everyone who has shown such an interest and enthusiasm for this book. Thanks to my colleagues at the University of Falmouth; my family and all the friends who have shared their experiences and love for dogs. Thanks to the Cornwall Social Media group where I met Tom, our wildlife photographer. He was perfect for this project, and provided many of the images.

The inspiration for this book came from growing up with our family dogs: Monty, Laddie, Shep, and Maisie dog. I miss you all but you live on in this book.

Special thanks to Jude Brooks of Hubble & Hattie for understanding what I was trying to say about the synergy between the psychology of people and dogs, and encouraging me to write it down.

Foreword

Dogs have evolved to live with people over many centuries, and there is something very special about the bond that we have with them. Recent scientific research reveals that dogs actually can understand us: 'read our minds,' and form close and long-lasting bonds.

A dog's intuition and ability to tune into us is the result of many years of observing us, reading our actions and behaviours, and learning to anticipate what we are going to do, based on a huge wealth of stored information our 'dog detectives' compile on us.

Even before we are ourselves aware of our intentions, our dogs are one step ahead – our automatic thoughts and behaviours acting as cues to them – and this book includes some of the scientific findings that show this.

If we have such a huge impact on our dogs – without even realising it at times – is it possible that we can learn more about ourselves from our dogs? Studies show that dogs can be aware of illness in people – such as diabetes and cancer – before the individuals concerned are aware of this, and many touching stories exist about how devoted our dogs can be in times of personal distress, providing comfort and reassurance.

How our dogs behave often seems to mirror how we are feeling, even though we may not know ourselves how we are feeling! We lead busy, hectic lives, often juggling jobs, relationships, financial issues, family commitments, career decisions, imbalances in work/life management, health concerns, losses, and continuing changes, all of which impact on our own wellbeing. As a result, we may feel anxious, sad, angry, fearful, guilty, tired, depressed, heartbroken or lonely, as well as many other negative emotions. Coping strategies can help, but throughout it all, our canine friend observes what is happening ... and stores it away in his 'reference library.'

Remember –
- if you become anxious, your dog will notice
- if you are feeling insecure, your dog will feel this
- if you experience an underlying sense of fear these 'vibes' are picked up by your dog

Equally, if you are calm and generally happy, your dog detective will pick up on this, too, and go about his life free from worry: playful, curious, energetic, affectionate, loyal, strong and busy – just being a dog, in other words. These types of behaviours are your dog's wellbeing barometer.

Of course, we may see behaviours in our dogs that cause us concern: incessant barking, destruction of furniture, aggression, fearfulness, skittishness. In order for dogs to live harmoniously with humans, we need them to behave in ways that we find acceptable. As our lives go through transitions and ups and downs, so, too, do those of our dogs.

Living with a dog is a two-way process, and this book aims to encourage owners to take some time to look at the psychological nature of the relationship they have with their dog. Dogs have evolved to become expert observers of people, and can reveal a lot about us if we know what to look for.

A dog may be man's best friend, but first we must try to be our own best friend. If we have chosen to share our life with a dog, we have a responsibility to take the best care we can of him or her: we also have a responsibility to take care of ourselves, too.

This book will help smooth the way as we grow even closer to our four-legged friend.

How will this book help my dog?
More relaxed and happy dogs live with more relaxed and happy dog owners.

How will this book help me?
It will enable you to recognise when you are experiencing negative thoughts, and perhaps struggling with life.

How will this book help me to live a happy life with my dog?
You will learn how to change your negative feelings into less unsettling thoughts, which your dog will pick up on, and, in turn, feel more secure and settled.

A happier you, a happier dog.

Remember –
- you cannot fool your dog
- he knows when you are struggling
- he can read you like a book!

In summary, this book will help you to appreciate and recognise how your thoughts and feelings impact on your dog. It will provide ways to recognise when you are struggling, and provide you with a CBT 'toolkit' to help deal with unsettling emotions. Paying closer attention to yourself will be a great investment in the relationship you have with your dog.

Introduction
If you have acquired this book it is highly likely that you have a dog or dogs in your life.

Perhaps you are the owner of a dog, or maybe you look after dogs when their owners are at work or on holiday?

You may volunteer for a charity that helps people who cannot look after their dog.

You might work with dogs, or maybe train dogs for others to share their lives with.

Whatever the situation, living or working with dogs means that you contribute to the bond that our dogs form with us, and this book is about the relationship between you and your dog.

Why is this book needed?
I am a dog lover, and I'm also a psychologist. I have been a psychologist for over thirty years, but I have shared my life with dogs for much longer.

Working as a psychotherapist; developing an understanding of people who struggle with life, I have been trained to use techniques and practical strategies that I teach my clients, to help reduce the unsettling and distressing emotions they may sometimes experience.

This specific form of therapy – Cognitive Behavioural Therapy (CBT) – is well known, and very popular internationally. I was just about to have my first book on CBT published when I met the publisher of the Hubble & Hattie books at a book fair. I loved her display of books about animals, and we got chatting, discussing how, sometimes, when our dog's behaviour changes, and they seem to become unsettled, it is often the case that their owners are also experiencing changes, and the anxieties and fears that this can create are transmitted to their sensitive dogs, who very often mirror this emotional state.

I suggested maybe I should write a book which could provide dog owners with some basic CBT techniques they could use themselves, which would, in turn, benefit the dogs. Using the knowledge I had gained from working with human psychology could help dogs who seemed unsettled and engaged in unhelpful behaviours.

What an exciting possibility: a natural way to combine my love of working with people with my love of working with dogs. It was as a result of that meeting that this book – maybe the first of its kind – came about.

Your dog and you – understanding the canine psyche

What is the canine psyche?

The word 'psyche' comes from the Greek word for soul, but, over the years, the term has become used in connection with the study of the human 'mind,' which includes the way we think: consciously, in that we are aware of our thoughts, and unconsciously, when thoughts are subliminal and involuntary.

In this book, we will consider what our dogs might be 'thinking,' or at least how they are processing information around them, and what effect this has on them. One way we can do this is to observe changes in their behaviour.

Research now uses clever scientific machines that can measure changes in brain patterns, allowing us to understand how different things affect the canine brain. Studies of the various types of barks and whines that dogs make also seem to indicate that the animals are processing the environment in different ways, according to what is happening.

We know that dogs respond in different ways to people expressing emotion, such as crying, or anxiety displays, and, in fact, empathise when this happens, showing concern.

What does this book offer that others don't?

Well, for a start, the book looks at things from your dog's perspective, and will explain and allow you to understand why your dog behaves as he or she sometimes does.

Rather than simply training our dogs to adapt and modify their behaviour to do what we want, let us consider how the world may look through their eyes ...

What *is* psychology?

In a nutshell, psychology is the scientific study of the mind and its functions, especially those affecting behaviour in a given context. This definition was used to refer to the human mind only, originally, but has changed over the years, and there are those who consider psychology to be the study of human and animal behaviour, with the aim of trying to understand why living animals behave in the ways they do.

Previously, the only way to study how animals behaved was to watch them over time, and make deductions from what was seen. We understand about their basic needs for food and shelter, and looking after their physical care, but we may also find it helpful to know about their natural behaviour, too, because if we can understand why our dogs behave in certain ways, this can help us in choosing the best ways to train them to live harmoniously with people. More recently, there have been advances in science that can help us learn more about a dog's mind. For example, we can use MRI scanners to actually see what changes occur in the brains of animals in response to changes in their environment.

A study on dogs in Hungary, led by Doctor Andics, revealed that dogs' brains react to voices in the same way that our brains do. The dogs had 12 preparatory sessions, followed by seven sessions of getting used to being in the scanner room. Using positive reinforcement, the eleven dogs who took part in the study were trained to lie still for up to eight minutes when in the scanner machine.

Researchers played two hundred different sounds to the dogs and the 22 humans who also took part whilst they were in the MRI scanner. These sounds ranged from environmental noises (car engines, whistles), to human sounds (but not words), and dog sounds.

Different breeds of dog can exhibit very individual traits and characteristics ... just as people do.

Your dog and you – understanding the canine psyche

Researchers discovered that the same area of the brain – that part at the front of the temporal lobe – responded in both dogs and humans to the sound of human voices. Researchers were surprised to discover that this area actually does exist in the canine brain, and reported that it was the first time they had seen this at all in a non-primate.

Emotional sounds such as crying and laughter also elicited a similar response in both species, in an area of the brain near the primary auditory cortex. And likewise, whimpering and angry barking produced similar results, although much more noticeably in the dogs' brains.

Dr Andics said: "We think dogs and humans have a very similar mechanism for processing emotional information. We know very well that dogs are very good at tuning into the feelings of their owners, and we know a good dog owner can detect emotional changes in his dog – but we now begin to understand why this can be."

Professor Sophie Scott, from the Institute of Cognitive Neuroscience at University College London, said that the study showed evidence of responses to emotional sounds that humans make, such as crying and laughing, sounds that are more like the calls that animals make.

The next stage in the research will be to determine if there is evidence that dogs can sense and understand the different words in their owner's speech.

What affects canine psychology?

All animal bodies – including our own – are made up of bones, tissues, blood and hormones, and development of an animal from conception to birth, and eventual growth into an adult, is affected by many different factors, two of which, in particular, affect the way an animal develops and behaves –

- nature – the genetic influence, which determines breed, shape, size, eye colour, hair, temperament
- nurture – environmental influences, such as parents, place of birth, surroundings, food supply

The combination of nature and nurture help make up an animal's psyche, and in both humans and canines, this comprises four elements – thinking, behaviour, personality, and emotions.

When did the study of psychology begin?

Throughout history, mankind has sought to discover and understand the reasons why we behave in certain ways: the Ancient Greeks, for example, were great scholars who were interested in the study of behaviour and the human mind.

Although there is evidence that dogs have been kept by humans for many centuries, studies of behaviour and the canine mind have only been conducted in the last century. Traditionally, dogs were highly regarded as loyal companions. The Greek philosopher Plato was very fond of dogs, and referred to the dog as a "lover of learning." Socrates considered the dog to be a "true philosopher."

Throughout history, many – including Plato – have considered dogs to be not only loyal, but also intelligent and thoughtful creatures: a concept borne out by the obvious comprehension in the eyes of these dogs.

Many stories tell of enduring characteristics, such as loyalty, evident in the psychological make-up of dogs, and the one that follows dates back to ancient Greece.

Argos belonged to King Odysseus, who lived on a Greek island. Odysseus went travelling, and was absent for twenty years. Upon his return, he discovered that his wife, Penelope, was being harassed by men who, assuming Odysseus to be dead, wanted to marry her. In order to secretly enter his house and spring a surprise attack on the would-be suitors, Odysseus disguised himself as a beggar. En route to his home, he came across Argos – now old and neglected – who, nevertheless, recognised his master and had just enough strength to drop his ears and wag his tail. Odysseus is unable to return the greeting of his beloved dog for fear of being recognised, and is forced to ignore Argos who, having seen his master once more, dies.

This is a lovely tale of a dog's loyalty to his master, even if he is no longer there; a theme of doggy devotion and constancy that can be found throughout history.

Another – well-known and true – example of canine allegiance is that of Greyfriars Bobby. This little dog was a Skye Terrier, whose master was John Gray – more commonly known as Auld Jock. John was a policeman, or 'Bobby,' as they were called in those days, in mid-1800s Scotland. A requirement of John's job was that he have a 'watch dog,' and he chose the little Skye Terrier when the dog was six months old. The pair were inseparable for the five years that John served as an officer in Edinburgh.

In 1857, John became ill and, over the next few months, his condition worsened until he died on February 8, 1858. Bobby was there at the funeral, and stayed by the grave all night. The curator of the churchyard, James Brown, removed Bobby, as he was not really allowed there, but Bobby returned for the next three days to sit on the grave. In the end, James took pity on the little dog and gave him some food. Bobby returned every day for the next fourteen years to sit on his master's grave.

The local community took Bobby to their hearts, and fed and looked out for him. He made many friends. Bobby died in 1872, and a memorial in his honour was commissioned by Baroness Angela Burdett-Coutts: a granite fountain with a statue of Bobby on the top, which sits on the pavement near the churchyard at the top of Candlemaker Row in Edinburgh.

Stories such as that of Greyfriars Bobby, who spent 14 years guarding his owner's grave, demonstrate how deeply dogs can feel.

Tom Walters, who took some of the photos for this book, visited the statue in 2013, to capture the above image of Bobby.

The inscription reads: 'A tribute to the affectionate fidelity of Greyfriars Bobby. In 1858, this faithful dog followed the remains of his master to Greyfriars Churchyard, and lingered near the spot until his death in 1872, with permission erected by Baroness Burdett-Coutts.'

Psychology origins

Psychology is still a relatively new science, and its beginning is often associated with Freud, a Viennese doctor in the 1800s, who was interested in finding out why the behaviour of some was outside the boundaries of normal mental health. Freud devised many interesting theories about why some individuals became very anxious, and behaved in unusual and often debilitating ways, most of which related to development from early childhood. His theories were developed into a specialised form of treatment called psychoanalysis, which deals with the interaction between the conscious and unconscious mind.

Other researchers studied human behaviour, and came up with different theories and treatments. Other early theories in the 1950s gave rise to the term 'behaviourists,' who believed that behaviour is a direct result of the experiences encountered – a 'stimulus-response' – which can be changed and reprogrammed.

Still other psychologists considered that behaviour was not an automatic stimulus response, and that humans were capable of choosing how to respond to any given situation and stimuli. This, they called the 'cognitive' process, from the Latin word *cognoscere*: 'to think.'

Cognitive psychologists claimed that, sometimes, our thinking process becomes 'faulty,' and we begin to interpret the world in negative ways as a result, which can cause unsettling feelings and unhelpful behaviours. But in order to help people resume 'normal' behaviour, it is necessary for them to identify their 'faulty thinking,' and work to rationalise this, reducing the unsettling emotions and unhelpful behaviours they may have developed.

So, what has this to do with dogs ...?
As mentioned, the study of dog behaviour is an even newer science, but one which has become increasingly popular in the last fifty years or so. There were very few dog behaviourists in the 1950s, and, although many people owned dogs, it was unusual for them to seek the help of an animal behaviourist for their pet. Indeed, traditionally in the UK, dogs were often let out of the house in the morning to roam about and take themselves for walks.

Since that time, our pace of life has changed dramatically, and it's the case that, in many households, no one is home during the day. Despite this, however, the number of households that own dogs has risen sharply: 4.7 million in 1965; 7.4 million in 1990, and 8.5 million in 2013.

The Pet Food Manufacturers' Association (PFMA) has been collecting data on pet ownership since the mid-1960s, but explains that the method of collection used has changed since 1980, so direct comparisons cannot necessarily be made. The figures that follow are taken from these collections, however, and the PFMA estimates that, by 2013, 13 million (45 per cent) households in the UK kept pets, around 19 per cent of which were dogs.

Alongside the rise in dog ownership, there appears to have been a corresponding increase in the number of animal behaviourists, with particular emphasis on help with so-called 'problem' dogs.

'Problem' dogs
It's well known that owners sometimes struggle with how their pooches behave, and seek help from the growing number of canine behavioursts now in business.

Debate about how best to train or 'retrain' a dog out of his 'bad' behaviour is usually lively. Most systems tend to focus on the canine, and seek to change how a dog behaves using behaviour modification techniques. Many of these programmes can be very successful, as we have seen when behaviour modification is used with people.

Dogs are not people, however.

We have purposely domesticated dogs to live with us in our homes, rather than outside in a kennel. Our dogs are now usually much-loved companions, without the requirement for them to undertake any of the usual roles

Some dogs prove a challenge to new owners, and many behaviourists claim they can train a 'problem' dog out of this 'bad' behaviour.

they were intended for, such as hunting and guarding. How we view our dogs has had an impact on canine psychology.

Why do we keep dogs?

These days most people keep dogs as companions, and there are many positive aspects to sharing our lives with a canine friend. Dogs have evolved to live with humans over many centuries, and there is something very special about the bonds that have developed as a result. Recent scientific research confirms that the notion our dogs are able to understand us – 'read our minds' – has basis in fact.

Canine intuition and the ability to 'tune in' to us is the result of years of careful and constant observation on the part of dogs, as they 'read' our actions and behaviours, and learn to anticipate what we are going to do. Amassing a huge wealth of stored information – data banks – our 'dog detectives' have become experts on us, and, even before we are aware of our own intentions, can be one step ahead: our automatic (and sometimes unconscious) thoughts and behaviours providing cues for our intentions.

Canine empathy

During the writing of this book, talking with dog owners I was told many times how their dog 'knew' when they weren't feeling well or a little low, or when family members were about to arrive ... how do our dogs to this?

Can dogs empathise – can they understand and appreciate what it's like to feel as we do? Research has shown that this may well be the case, and evidence for this is included in the book.

If we have such a great impact on our dogs, without even realising it at times, perhaps we can learn more about ourselves from our dogs. Hard evidence exists that dogs can be aware of illness in people before they are themselves, and there are many touching stories about how devoted and attentive our dogs can be in times of personal distress, providing comfort and support in troubled times.

Our dogs' behaviour often seems to mirror how we are feeling – even though *we* may not know how we are feeling.

Canine cognition

It is becoming increasingly obvious that dogs possess an understanding of human behaviour that other animals do

Many people claim their dog understands their mood and behaviour, but can a dog *really* feel empathy for a person?

not, and centres of study are devoted to investigating and determining precisely *what* dogs can understand.

Some of these are known as 'cognition centres' because the aim is to investigate what processes are taking place in the dog's brain, and try to determine exactly how much they can understand.

Dog training

Training and modifying canine behaviour goes way back to recorded scientific experiments, such as Pavlov's dogs.

Pavlov, a Russian physiologist, observed that the dogs he and his assistants were working with salivated whenever one of his assistants entered the room. Because edible and non-edible items were used in the research, they had come to anticipate receiving food from the assistant, and salivated in expectation of this. This, Pavloc decided, was not the unconditioned, automatic physiological process of salivating in response to food, but a conditioned, learned response to the stimulus of seeing the assistant.

As an experiment, for a period Pavlov rang a bell whenever the dogs were given their food; then he simply rang the bell without providing any food. Not surprisingly, the dogs salivated upon hearing the bell, as they had come to associate the sound with food.

This is known as classical conditioning – the first systematic study of basic laws of learning/conditioning – which forms the foundation of what has become behavioural psychology: important today for use in behavioural modification, and mental health treatment.

How does this apply to people?

Classical conditioning has demonstrated that humans can be 'trained' in the same way as dogs. Interestingly, although behaviour modification is a method still used today for people, it was noticed that just because we change our behaviour, it doesn't mean that we also change how we actually think.

People may modify their behaviour in order to gain something they want, but it doesn't always follow that their way of thinking has changed. We have the ability to think about what is happening, and consciously choose what response to make; dogs don't have the intellect to be able to reason about their decisions. There is a stimulus, they react, they change their behaviour.

Dog training – and 'training' the person who looks after this dog – has become a large part of caring for a dog.

Your dog is watching you!

Our dogs watch us constantly, and have evolved to be successful at living with humans because of their ability to adapt and become part of our family, or social group. Their ancestors – wolves – possess an incredibly high level of intuition, with pack members apparently able to understand each other, and what each is about to do. The co-operation and co-ordination in wolf packs mean the units function exceptionally well, which increases their chance of survival. A wolf uses these highly refined senses to be constantly aware of his environment, receiving and processing information, and acting upon this, where necessary.

All five senses – sight, sound, smell, taste and touch – come into play when a wolf checks out his surroundings. Pack survival is paramount, and the highly developed social networking that wolves employ maximises the likelihood of a well-fed, secure and healthy unit.

From puppy to adult dog

PUPPIES AND PLAY

Young wolf cubs can be seen to engage in a lot of play. Their rough-and-tumble games are not discouraged by their mother, and this seemingly 'pointless' activity (in terms of providing for the welfare of the pack) often puzzled us. Of course, it has been realised now that play has a very important role in the development of physical and thinking skills in wolves: some researchers believe that, by the age of four weeks, a wolf cub will have established how he will communicate. Dependence on the mother gives way to increasing socialisation with siblings, and play is a very important part of learning these new skills. The wolf pup's development in physical, emotional, problem-solving, social bonding, co-operation, operation and territory marking terms is all learnt through early play.

As the cubs grow older their playful activities are channelled and guided toward more specific pack survival activities, such as hunting, stalking, learning how to kill prey, and how to bury and store food. An interesting fact is that studies of wolves show that, even when they grow older, they still enjoy play, and can be observed racing around in chasing games, hiding and pouncing on each other, and generally 'letting off steam,' just like children playing, having fun and socialising together. Often, people lose this ability to play and behave in a 'free child' state as they age and live as

Your dog probably notices more than you think: an innate requirement to ensure their survival means dogs are always on the alert.

adults, the pressures and responsibilities of surviving in the world taking precedence over time out to run and jump.

Children and puppies

Children can play an interesting and important role – for both dog and child – in living with dogs, as physical and emotional wellbeing can certainly be enhanced by play,

Puppies, just like their wild wolf cub counterparts, use play to develop life skills, and to establish their place in the social group.

fun, and laughter. The connection between living with and caring for a dog, and the subsequent effect on our emotional and physiological states is discussed later in the book.

Hierarchy

Wolf pack hierarchy is something we are still trying to properly understand. However, we do know that the 'alpha pair' (male and female) lead and guide the pack, and are responsible for producing and raising the young (most packs contain only one breeding pair). At the other end of the scale is the 'omega' – the lowest ranking wolf in the pack, who can be male or female. In-between come offspring of varying ages, though generally below the age of sexual maturity.

The situation for family dogs is similar, depending on how many members comprise the social group.

Your life with your dog

How your dog develops whilst living with you will have a significant influence on the relationship between you. The first four weeks of a dog's life with his mother are very important, as a lot of learning goes on during that period.

Social interaction begins with the mother before the pup's eyes are open, with smell and touch the main sensory inputs. By four weeks old, the ears are open, and the pup is ready to receive input from all five senses.

During the next period of socialisation, hierarchy – the pup's place in the group – is established via playing and rough-and-tumble activities with his siblings. The UK charity Dogs Trust says that a pup is not ready to leave his mother before eight weeks of age, and the UK Dog Advisory Council publishes a seven-point guide about choosing a dog –

- choosing a breed
- choosing the right dog: exercise requirements of puppies and dogs
- choosing the right dog: costs of dog ownership
- rehomed dogs
- finding a trustworthy breeder

- socialisation of puppies
- puppy contacts

– which is extremely useful if you are thinking about getting a dog, because the reasons why you want one, what breed you choose, and when you decide to acquire a dog are all significant factors in influencing the relationship you will have with your dog.

Looking at your own situation and motivation for choosing to share your life with a dog – as well as your personality and attitude about looking after and training a dog – are also key elements to building a successful and mutually fulfilling relationship with your canine companion.

Your psychology with your dog
Points to consider –

- what motivated you to get a dog?
- which type of dog did you choose?
- why did you choose that particular dog?
- what is it about your dog that you especially like?
- is the look of your dog important to you?
- is the personality of your dog important to you?
- how do you think your dog influences/affects your life?
- what do you think your dog says about you?

Answering the above points should demonstrate how your dog's attitude, character and temperament can be shaped by your influence and attitude.

Considerations for having a dog
Later in this book we look specifically at you and your personality and situation, but in the last section of this chapter we take some time to consider why you have/may want a dog. The following statistics and advice from various dog organisations – as well as general points to consider – may help in this respect –

Giles Webber, Operations Director of Dogs Trust, says: "Everyone in the family should talk about why they want a dog. The whole family must be committed." He quotes statistics which show that 11,986 stray and abandoned dogs were picked up in the UK in 2013 – a staggering 307 *a day*.

Carly Whyborn of Battersea Dogs Home asks potential dog owners to consider what their working hours are each day; what space is available in their home, and whether their holiday plans would include a dog: "Can you cope with up to fifteen years of commitment? she asks.

And, make no mistake, taking on a dog is usually a long term commitment, during which time family dynamics can – and probably will – change. Those very same youngsters who pleaded to have a dog, and promised faithfully to walk him or her, look after them, and dogsit to allow you, their parents, to go on holiday, etc, grow into teenagers, whose sleep patterns may not be conducive to early morning walks, and young adults whose social lives may require freedom to roam. I really wanted a dog for our family, but, as I worked full-time, I waited until the children were old enough to care for a dog, and take him or her for walks.

Getting a large, boisterous dog may not be the best idea if you're older, and not able to manage the exercise requirements of such an animal.

When considering getting a dog, and what type of dog to get, Giles Webber recommends putting the dog first: "People need to imagine themselves with four legs and a tail," he says, and poses the question: would they like to be their own dog?

Be honest about the potential problems of owning a dog. If, because of lifestyle/commitments/work, a dog will be 'regularly marginalised' – ie be a secondary consideration, possibly spending long hours home alone, then maybe owning a dog is not a good idea for you.

Also take into account the cost of food, insurance, veterinary bills, micro-chipping, pet passport, boarding kennels, and time necessary for regular exercise and play.

"Consistency is everything to dogs. Most want to please you, but they need direction," says Carly Whyborn.

If you love being with dogs and caring for them, but aren't in a position to offer one a good home, another possibility is to dog-share. A great scheme called 'Borrow my doggy' was set up in 2012 by Rikke Rosenlund. The idea is that dog owners and their dogs are matched to individuals who don't have a dog, taking into account breed, temperament, family situation, and free time. Rikke explains: "There is no need for a dog to be home alone, or for an owner to pay for a dog walker when there are loads of people who adore dogs, but can't have one of their own." By October 2013, after setting up the website www.borrowmydoggy.com, thousands of members had been signed up. Doggy borrowers and lenders range from a student away from home who really missed being around her dog to an older man who was no longer able to take his dog on a longer walk.

"When we started out we asked our members why they had signed up and the letters I received in response made me cry," says Rikke in an article that appeared in a UK newspaper.

Conclusion

In this first chapter we have looked at some aspects of how the dog's mind has developed, and how understanding what is going on from a canine point of view can help us to understand them. This is as useful for those who already share their lives with dogs as the prospective dog owner, because dogs constantly change and develop, just as we do. The bonds we form with our four-legged friends are dynamic, so perhaps we should rethink that old expression: 'You can't teach an old dog new tricks' because you evidently can. Evidence seems to show that dogs can be just as flexible and receptive to re-training, different people and surroundings as us.

In the next chapter we look at ourselves more closely; becoming aware of the significant influence we have on our dog's behaviour.

The way you behave around your dog can directly influence the relationship you have, and we find out more in Chapter 2 ...

It's no easy task trying to deduce what a dog may be thinking, and often we attribute human characteristics – such as happiness, guilt, sadness – to them, though cannot actually know whether this is how our dog is really feeling.

Recent research has used sophisticated scientific devices to track activity in the canine brain in response to particular stimulating situations – different sounds have been played, for example – and the results of these studies are helping us understand the 'canine psyche.'

Of course, our own behaviour may affect how our dogs think and feel. For example, if we are feeling very anxious or upset, we may see a corresponding change in our dog's behaviour. Our demeanour can impact hugely on our dogs, whose own behaviour may be affected as a result.

Therapy

When we're upset, we sometimes seek help to manage our unsettled emotions, which can take the form of counselling or therapy.

Many different types of therapy exist, one of which is CBT, in which I am trained, and believe can be very helpful at difficult times. This book was going to be called *CBT for dogs*, but, as we were not sure that everyone would know what CBT stands for, we decided against it. Some helpful friends suggested that CBT stood for 'Cold Beer Therapy,' or 'Compulsory Basic Training' (for motorcyclists), but in this instance, we mean Cognitive Behavioural Therapy, of course: an increasingly well known, practical form of psychotherapy, designed to help those struggling with emotional ups and downs in their lives.

The original intended title also suggested that CBT could be taught to dogs, when, really, the idea is that we can use this therapy to help ourselves through tricky spots,

which will, in turn, benefit our dogs if we feel calm and secure in our emotions.

Perhaps one day it *will* be possible to teach dogs CBT – in which case, maybe it should be called *Canine Behavioural Therapy* ...?

Wolves and dogs

In recent years there has been a lot of interest in research into how dogs have developed from our working partners when we were hunter gatherers, to our beloved companions and 'man's best friend' in today's society.

The evolution and domestication of dogs is an amazing story.

A theory exists that, as man and wolf are both meat eaters and daytime hunters, they had a common interest and objective. Wolves would wait on the periphery of camps to scavenge for the bones and scraps of meat the villagers discarded. As time went by, they became less fearful of man, and began to live in closer proximity to him, with the tamer wolves eventually residing in the camps alongside man.

Although dogs have not always been around, geneticists have ascertained that the dog's closest relative is the wolf. Some dog breeds look very different to wolves, and many scientists agree this is because those wolves who associated with man became domesticated, and underwent genetic changes that gave them a greater chance of surviving in a human environment. They also deduced that those dogs who lived in human society for a long time developed different thinking skills from wolves: learning to read and interpret hand signals – for example, where food was hidden – and so their brains developed in different ways to those of the wolves.

Dr Adam Miklosi, biologist and founder of the

When contemplating dog training, it's important to consider our approach to and part in this, also, in order to more fully understand our dogs – including, perhaps, his history and evolution.

Family Dog Research Project, which is widely viewed as the formative research centre for dog cognition, has closely studied the wolf and dog link, and has discovered differences in –

- thinking skills
- attachment behaviours
- communication skills
- emotional behaviour (for example, wolves use

barking as a warning only, but dogs bark for many different reasons, and have different types of bark, which owners can often distinguish and identify the reasons for, such as anxiety, fear, playing, 'welcome home' barking, and separation anxiety)
- personality (wolves would not adapt their personality to suit people, but dogs can be seen to adapt and change to do so. Dr Miklosi calls this 'mutual aclimatisation.' He is also interested in

determining whether dogs and people actually affect and influence each others' personalities)

Siberian Silver Fox experiment

Russian biologist Dmitry Belyaev felt that genetics were at the heart of how such an incredible diversity of dogs had arisen from their wolf ancestors, and with fellow Russian, researcher Dr Ludmilla Trut, carried out a remarkable study.

Deciding to use foxes as the base species of his study, to see what they could tell us about the domestication of dogs, Dr Trut travelled to Soviet fur farms and selected the calmest foxes she could find, to serve as the base population for Belyaev's experiment in Siberia. After eight generations of breeding for tame characteristics, Trut noted there were some real changes in the foxes, which had begun to occur, in fact, after only three generations of breeding for this trait.

She wondered, were the foxes tamer because they lived in close proximity to people?

To test this, Trut allowed another group of foxes with aggressive characteristics to breed. She then gave some of the resulting fox cubs to the tame mothers to nurse, to try and establish whether a tame vixen could influence the behaviour of the aggressive cub, making it tamer.

Result – it made no difference. Nature won out over nurture and the aggressive cub remained aggressive. Trut concluded that genetics were responsible for the difference between tame and aggressive foxes, and called this a 'gene for tameness.'

Living with wolf cubs

Researchers in the US wanted to determine whether the way that dogs are raised in the home is the reason why they are so different from wolves.

What would happen, they wondered, if five wolf cubs lived with them from the day they were born?

To provide a control for the study, initially, five day-old puppies were raised in a domesticated setting, after which five day-old wolf cubs were introduced into the same

Dog breeders will selectively breed those dogs with the most desirable characteristics: the Labrador, for example, is known for his easy temperament and amenability, and is hard-working, loyal and affectionate – a perfect family dog in many respects.

Opposite: Study findings confirm that a dog's natural instincts cannot be altered, but that selective breeding ensures that the most favourable characteristics are brought to the fore ...

... and can also influence and dictate physical appearance to better appeal to people (smaller and cuter, perhaps).

Opposite: Many people find that a puppy or dog elicits in them the same protective and nurturing feelings that babies and children do.

setting, and taken everywhere with the researchers; thus forming a strong relationship.

By the time the cubs were eight weeks old, however, changes in their behaviour were noted. Although the cubs were still interested in the humans, they were not as co-operative as they had been, and existing battles between the cubs and a couple of their carers became worse, to the extent that there was constant conflict. The cubs seemed to want to destroy everything, and, after four months, had to be returned to the reserve as they had become unmanageable.

The researchers concluded that the difference between dogs and wolves must be due to the way that dogs have been bred, bringing out desirable characteristics whilst reducing those less favourable, which seems to reinforce the findings of the study done with the Silver Foxes.

Selectively breeding tame wolves with other tame wolves eventually produces animals that are tame enough to live alongside people.

But what about the physical changes that occurred during the wolf-to-dog evolution? This part is so interesting – and clever!

As the tame wolves evolved, it was noticed that they began to show what seemed like affection towards humans. They were also more co-operative, and, in fact, made eye contact and watched the people in their environment, both of which can establish strong bonds between animals.

As soon as they began to crawl, the cubs would breathe heavily (like panting), wag their tails, and howl: behaviour which humans find appealing.

Physical changes, such as tail length, coat composition, floppy rather than erect ears, and shorter limbs became evident in some wolves (Dr Trut had noticed that the tame Silver Fox cubs were beginning to look like dogs as she knew them).

Associate professor of evolutionary anthropology at Duke University in North Carolina, and member of the Center for Cognitive Neuroscience, animal behaviourist Professor Brian Hare has made many studies of dogs,

and noted that: "If we select for tameness, then physical characteristics will change also."

He further observed that human beings favour 'juvenile' characteristics in dogs – the 'cute factor' – and dogs appeared to be evolving to have these characteristics, thereby increasing their appeal to us, as demonstrated by the words we use to describe them: cute; adorable; funny. He reports that breeding dogs to look this way reveals a lot about us, and the traits we like our dogs to have.

Another researcher, internationally recognized, prize-winning neuroscientist Professor Morten Kringelbach, from the University of Oxford, says that dogs fulfil our need to nurture, which is a very strong basic instinct. He believes we are hard-wired to respond favourably to infant-like features such as a large forehead, big eyes, and big ears.

Kringelbach carried out a study which tracked the human response to pictures of infants and adults, and noted that, within one seventh of a second, activity occurred in the brain just above the frontal cortex when a picture of a baby's face was presented, but not when an adult face was shown, demonstrating that this was an emotional response. He purports that we respond similarly to dogs, "... as though they are children. Their features elicit similar emotional responses."

The other interesting aspect of this evolution of features and selective breeding is that it ensures survival of the canine species; Professor Kringelbach even suggests that we may be moving away from having children to having pets!

Doggy emotions

Do our dogs experience emotions? Most dog owners would probably say yes, they do. Upon making it known that I was writing this book, people were keen to share stories about their dogs, many of which were to do with how their dog was 'feeling.' Some told me how upset their dogs get when they leave the house without them; some claimed to know if their dog had misbehaved whilst they were out as he or she would look 'guilty' upon their return. Others might say how jealous their dog became if they interacted with others or patted someone else's dog, or become very protective or particularly attentive if they felt ill or unhappy.

In the past, there was some scepticism about interpreting some canine behaviours as emotional responses, but, more recently, scientific research has shown that dogs do experience emotions. I have included mention of some of these studies in this chapter as science shows there has to be a link between cognitive processing (thinking) and a behaviour in order to demonstrate some of these responses.

The previously-mentioned Dr Adam Miklosi in Hungary wanted to investigate how man's best friend has become so involved in our everyday lives. Previous research looking at how animals communicate with each other in social groups involved animals such as monkeys, dolphins and ants, but Dr Miklosi wanted his research to concentrate on dogs. Until then, most scientists were of the opinion that, because dogs were domesticated and close to humans, it would be difficult to study them objectively. Now, however, scientists are considering the possibility that studying how dogs have developed their clever 'thinking' skills might help us better understand not just dogs, but ourselves as well!

Consequently, the first dog cognition conference was held in 2009, and the second in July the following year at the Clever Dog Lab and Wolf Science Centre (WSC) at the University of Vienna. The main focus of the work at the Clever Dog Lab is to try and understand the cognitive and emotional abilities of domestic dogs, particularly in relation to their ancestor, the wolf, by undertaking studies not only on early canine development, but in old age, too.

The final area researchers are especially interested in is how dogs develop "specific mechanisms of imitation and empathy, as well as co-operation, and gestural and vocal communication."

Doggy happiness

Can you tell if your dog is happy? For that matter, can dogs *feel* happiness, or is that simply wishful thinking on our part?

Dr Tina Bloom, a psychologist from Walden University in Minneapolis, writes about how most people – even those who don't interact with dogs much – say they can tell if a dog is happy, sad, surprised or frightened.

She says: "Canids are highly social, and communicate using an abundant array of facial expressions." Her assertion

Therapy dogs are now commonly used in hospitals and the like, helping to provide emotional support to people of all ages.

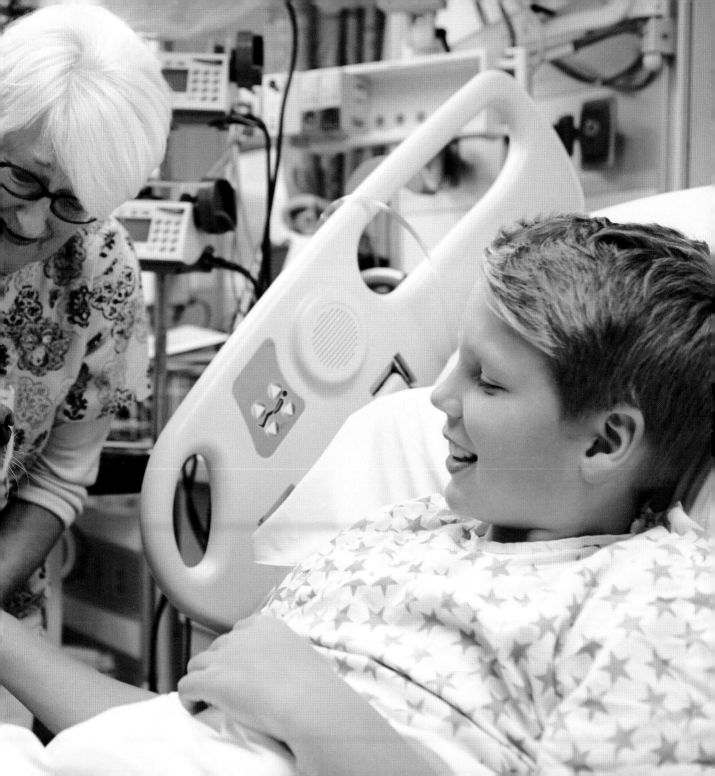

is interesting to consider, and perhaps compare with how adept humans are at reading, say, the facial expressions of other animals. There is no doubt that people and dogs have developed a special closeness over thousands of years, resulting in a natural empathy.

We generally judge whether or not a dog is happy by a wagging tail and facial expressions. Animal behaviour specialists have found that dog emotions are really shown in their faces. Using high speed cameras, they tracked facial movement in dogs in different situations – for example, when they were reunited with their owners or a stranger. Coloured tags were used to help track movement, and attention focused on eyebrows, ear movement, and movement in the side of the face.

What they discovered was that dogs tended to raise their eyebrows for about half a second after meeting up with their owner, but if it was someone they didn't know, they would move their left ear back a little. If the dogs were shown an object that they didn't like they would move their right ear in the same way.

Another fascinating finding was that dogs seem to show a 'left gaze bias', which, apparently, has previously only been known in humans. In people, the left side of our brain controls the right side of the body, and the right side of the brain controls the left. The left side of the brain controls how we show emotion, which is displayed on the right side of our face. The left side of the face shows almost no emotion because the right side of the brain has a different function (doesn't control our emotional state of mind).

Looking at the right side of the face is called 'left gaze bias', and no other species does this when meeting one of its own kind: dogs only do this when they look at a person, and are therefore unique among the animal kingdom as the only species that can see and understand our emotions by looking at our face. Check out the website of animal behaviour therapist Sarah Whitehead, where you can take her quiz to establish if you can 'talk dog' (www.learnto talkdog.com).

Doggy jealousy

Recent research is showing that dogs do experience emotions similar to our own. For example, scientists in California have discovered that dogs show jealous behaviours. They set up situations where dogs could see their owners giving attention to quite realistic-looking stuffed toy dogs. The dogs would begin touching, nudging or pushing their owners when they showed affection to the toy dog, and sometimes they even snapped at the dummy dog.

We certainly found that, with our dog, Maisie, if anyone in our family went to give me a hug, she would jump up and bark, and try to get between us. I used to think it was a protective behaviour because maybe she thought I was being attacked (as the omega wolf will sometimes do to distract other pack members facing up to each other), but this recent research suggests it could be jealousy that makes her behave in this way.

The scientists studied thirty six dogs in their home environments, and the reactions were similar in most cases. They have concluded that jealousy is a complex emotion involving a third party, which was previously thought to be experienced only by humans, but that when dogs perceive that the special relationship with their owner is under threat, they will protest. In evolutionary terms, this shows that a dog is protecting his relationship with us in the interests of survival.

Doggy guilt

This is a hard emotion to prove in dogs. Certainly, our dogs do appear to show some behaviours which seem to indicate they are aware of having done something wrong, but we can't be sure if the dog links an anti-social act to the disapproval shown by their owners.

It is almost certainly the case that dogs will not understand the association between a transgression on their part, and being told off for it some time later (when you return home, say), as they will already have forgotten all about what it is they did, and will not understand why they are receiving a telling off. Some people think dogs 'act guilty' to reduce the time of negative interaction they receive, and that this could be a learned response.

Author and researcher in animal cognition Alexandra Horowitz found that dogs who had commited a transgression and those who hadn't were likely to exhibit similar guilty behaviours in response to their owner's disapproval, reinforcing the view that this is a learned response. However, those dogs who had actually done something they had been told not to (such as eating food when their owners were not there) were likely to act guilty for longer.

I remember coming downstairs the morning after I had been away for a week, to find my dog sitting under the table, and not coming out to greet me. I discovered that, during the night, she had pulled my sunglasses and mobile phone off the table, breaking the glasses and leaving teeth marks in the phone. The dining room table had been covered in the usual family junk, but it was only my belongings she had interfered with. Was sitting under the table and not coming out signs of 'guilty' behaviour? It was certainly out of character for her to do this.

It is difficult to really understand what a dog may be thinking, but being aware of your dog's usual behaviour, and identifying reaction patterns can give some insight into how your dog is processessing the world around him or her.

Our relationship with our dogs is very special and unique; the better we can understand them, the better we can build loving, healthy relationships with them.

The next chapter looks at the psychology of people, and how this can influence our dog's behaviour.

Visit Hubble and Hattie on the web: www.hubbleandhattie.com & www.hubbleandhattieblogspot.co.uk • Details of all books • Special offers • Newsletter • New book news

27

First, learn about yourself

This chapter provides a chance for you to consider your own psychological make-up. We are all unique individuals, comprised of different behaviours and experiences, which interact to make us who we are. Personality is influenced by many factors.

What sort of personality do you have: what are you *like*? This chapter encourages you to look at yourself, to determine how much you understand about your personality and psychological profile.

Your innate personality traits define who you are; your dog can sense this so it's important to understand who you are and how it can influence your relationship.

Appreciating that your dog is constantly watching you, making inferences about you and behaving accordingly, having some self-knowledge could be helpful.

A little test
Which of these two lists best describes you and your preferences?

A

- I prefer to have lots of quiet time
- When I am alone, I get my best ideas and also a sense of being myself
- I do attend busy social events, but feel a little unsettled when there
- I can talk to people and enjoy short periods of socialising
- I like to spend quiet time thinking and reflecting on life
- I enjoy the peace and quiet to be found in nature
- I like to have friends who I share an interest with – such as dog walking

B

- I like to be on the go, with lots of activities planned
- I can get bored easily
- I like to be out and about; active and busy
- I tend to lose energy when I am on my own
- I get distracted quite easily
- I tend to think as I speak
- I actually think better when I am talking

It's obvious from the above lists that A describes a more introverted personality and B a more extrovert one. The characteristics described in each are only part of a personality, of course, and some researchers believe that our basic personality is genetically influenced, and remains essentially the same throughout life.

Just as being in the wrong job or relationship for our personality can create tension and leave us feeling unsettled, so, too, can choosing a dog whose personality is at odds with our preferred state of being.

It is thought that personality characteristics fall into five categories –

- extroversion/introversion – as the foregoing two lists
- agreeableness – friendly, outgoing/reserved, reticent
- conscientiousness – pays attention to detail, cautious, measured/impulsive, easygoing, less attention to detail
- openness – cautious, deductive, holding back/ outgoing, divergent thinking, creative
- neuroticism – sensitive, cautious, nervous/ confident, secure, takes risks

Obviously, these characteristics and personality traits are found along a continuum, and research seems to indicate that 40-50 per cent are genetically pre-determined.

It is possible to change and refine personality traits, but this is harder than if changes occur naturally. For example, naturally quiet, reserved people may decide they need to be more assertive with their dog, but they will find trying this difficult and uncomfortable as it goes against their natural inclinations. Trying to be someone you are not is pointless, anyhow, as your dog will soon pick up that you are not being genuine. It is neither necessary nor desirable to shout at a dog or get angry with him, as he will feel threatened, and may respond by barking and becoming excited or, conversely, withdrawing and cowering.

Psychology studies show very clearly that people are not able to learn when they are frightened, and can develop coping strategies when they feel under attack that may not be healthy.

The same is true of dogs. When a dog perceives that he is under attack, this elicits the 'fight or flight' response (explained more fully in chapter eight, when we look at signs of distress in dogs), and also feelings of insecurity: not a good basis for a happy, harmonious dog – or person.

Know yourself
It can be helpful to have some understanding of your basic personality and preferences if choosing a dog to share your life with: finding a dog who will complement two people in a relationship can in itself be a point of conflict.

Family life produces its own particular joys and struggles: a new baby in a family with a dog who is well established can have a profound effect on the animal,

continued page 32

A clash of personalities can make training challenging.

A dog will understand and welcome firm commands that give his actions purpose, but shouting and anger can have the opposite effect.

Ensuring that the personalities of you and your dog are a good match – before making a commitment to look after them – is an important step.

and you may find yourself under stress because of divided loyalties between family members and your canine companion, so it's important to know your boundaries and tolerance limits.

Well-known British broadcaster and journalist Mariella Frostrup wrote an article about how, when their two dogs had to be cared for by friends and neighbours for two nights a week, the dogs' toilet habits changed for the worse, and her "two perfectly house-trained dogs who lived happily in our apartment in London," began to leave unwelcome deposits in the morning.

Their vet thought that the change in routine and disruption the dogs experienced had probably caused the change in bowel habits, and advised they concentrate on a strict routine, with meals at exactly the same times every day and regular walks morning and night, to "help diminish their [the dogs'] agitation." Mariella explains they took his advice and have not had a problem since.

Creating harmony

It can be a good idea to make a 'wish list' of how you would like your dog to be, and also take into account what you *think* you are like. Under the headings: What I think I am like/ How I would like my dog to be, add as much detail for each.

Now, under the headings: What I actually *am* like/ What my dog actually *is* like, again, add as many points as you can for each heading.

How do the two lists compare, and are there areas that need addressing?

Next, consider who else forms part of your social group, and write down their characteristics/personalities.

How your dog and the other members of your social group interact can be very different in each case; maybe you are already aware of these differences, and have noticed how your dog's behaviour changes accordingly. It is important to try and establish continuity and consistency of care and training for your dog, who will be more settled if there is harmony between and with all of his care-givers.

Everyone within your close community of friends and family will interact with your dog to some degree.

Homes4Hounds

A dog-minding service – Homes4Hounds – for people going away or on holiday was an excellent example of how to get the dynamics right, as I personally experienced. This was a new type of service for people who wanted their dog to be looked after, but not necessarily on a boarding kennel basis.

Our dog, Maisie, was seven years old when it dawned on us that we had no one in the family who could look after her if we wanted to go away. I discovered that a lovely lady called Michelle had set up the business Homes4Hounds, which would find the best temporary home for your dog in the homes of other people.

Michelle came round to meet Maisie, and discuss her exercise habits, eating patterns, likes/dislikes, experience of living with children, and overall personality, then asked what sort of home we would prefer our dog to be placed in.

Your dog and you – understanding the canine psyche

With pictures she had taken of Maisie, Michelle looked at her register of local, council-approved dog carers (who were all insured, too), in order to find the best match.

Next came a visit to the prospective dog-sitter, and some time together. Afterwards, Michelle asked both parties whether they felt comfortable with the placement, and, if we were, a booking would be made.

This arrangement worked very well for the next eight years, and Maisie stayed with five different families during that time. The home she went to mostly for the last three years of her life became her real 'home-from-home,' and she would prick up her ears when we were a mile or so from the house, positively bounding up the drive to greet the lovely couple who looked after her in our absence.

Opposite & above: Interaction with all those your dog comes into contact with should be consistent to avoid confusing him, which could give rise to unwanted reactions and behaviours.

Homes4Hounds seemed to me a very well thought out scheme, which minimised any potential separation anxiety and stress due to radical changes in the dog's environment and routine.

Empathetic understanding

The emphasis in this chapter is on taking time to consider yourself, and your dog's point of view. Trying to understand someone else's point of view, as if you were standing in their shoes, is being empathetic. Let's take time, then, to try and understand our dog's point of view, as if we were him or her, by thinking about the following points –

- how do you think it feels for your dog to be part of your life and community?
- does everyone your dog comes into contact with treat him in the same way?
- is everyone 'on the same page' with regard to the boundaries you have set for your dog?
- what do you think might happen if your dog receives confusing signals from different members of his community?
- your dog has to be a mind-reader, but what if he gets it wrong?
- how will he feel if he can't work out how to

If welcoming a puppy into your home and your lives, it is important to be both mentally and physically prepared to consider another creature's needs. Even so, there may still be times when you are tested, and need to remain in control.

react, and is told off as a result? Could he become confused, possibly agitated, and quite likely stressed?

Thoroughly reflect on your answers to the above questions: they may well provide insight into how your dog fits into your life, and you into his. This could lead to you deciding that some changes are in order: changes that may benefit you both.

Having considered some of your own personality characteristics in this chapter, you may gain further insight into how special your life with our dog is; your preferences and interests interwoven into your life together.

Your partnership is unique: no one else in the world will experience exactly the same human-canine bond that you have. Your history with your dog may also have a significant influence on your relationship: at what age you both were when the relationship was formed, for example, and how long you have been together.

In the next chapter, we look at the likenesses that you and your dog may share, plus wider studies of other people and their dogs, and the similarities between them, not only in temperament but even physical characteristics! This latter aspect may be something you've not really noticed, but the photographs of pooches and their owners in the next chapter are enlightening in this respect.

Visit Hubble and Hattie on the web: www.hubbleandhattie.com & www.hubbleandhattieblogspot.co.uk • Details of all books • Special offers • Newsletter • New book news

37

In the previous chapter I encouraged you to look at yourself in order to identify aspects of your personality and psychological profile: in this chapter we move on to considering the relationship between you and your dog, and how you interact.

On the basis that your dog is constantly watching you, making inferences about you, and behaving accordingly, having some self-knowledge could be helpful to being more aware of how you might be influencing your dog's behaviour (which you may already be aware of to a degree).

We can often discover things about ourselves from other people and situations. When my son was sixteen and a college student on a media course, a particular photography assignment required him to take photographs outdoors, develop them, and come up with a theme.

He, like me, loves dogs, and chose our Border Collie, Maisie (then a pup) as his theme, deciding that he would visit the park to take pictures of dogs out on walks with their owners. Suggesting that, perhaps, people may be a little concerned about a lone teenager taking photos, Maisie and I went along, too.

Upon developing the images he had taken he made a remarkable discovery: there were striking physical similarities between the dogs and their owners, including me and Maisie, as I had a tendency to always wear black and white, and our dog was a black and white Border Collie! Consequent research into how human personality traits appeared to influence the type of dog they chose revealed some similarities in each case.

Hey, good looking!
Think about your physical appearance, and how you dress.

For example –

- what shape is your face?
- what size and shape is your nose?
- what colour is your hair?
- what shape are your eyes?
- is your hair long or short?
- does your hair frame your face?
- what are your ears like?
- which colours do you prefer to wear?

Psychology professor and well-known author Stanley Coren found in his research that women who wore their hair longer and partially covering their face tended to have a Beagle or Springer Spaniel, as they considered these dogs to be friendly, likeable, loyal, and intelligent.

It can be fun and illuminating to study previous photos of you both to check for resemblances, as well as considering how you currently look in relation to your dog.

Much research has been done on the likenesses that exist between couples, and a favourite psychology exercise for students was to cut in half photos of couples on their wedding day, mix them up, and then ask people to re-match up the pairs. The percentage of correct pairings was high. A psychologist – Bob Zajonc – discovered that, as couples age, they seem to look even more alike.

Researchers from the University of California in San Diego, Nicholas Christenfeld and Michael Roy, decided to develop a study that Stanley Coren had begun. Photographing 45 dogs (25 purebreds and 22 crossbreeds) separately from their owners, they asked 28 volunteers, with no previous knowledge of the dogs and individuals concerned, to pair owners and dogs.

The hypothesis that owners and their dogs look alike is difficult to prove scientifically,
but anecdotal evidence abounds within studies ...
and our daily lives (and overleaf).

Amazingly, in about two thirds of cases the volunteers were able to correctly pair the purebred dogs with their owners, apparently supporting the idea that people do tend to look like their dogs.

The study found that, when people choose a dog, at some level they are looking for one who resembles themselves, rather than a case of owners becoming more like their dogs over time. Six human characteristics were considered when comparing owners and dogs –

It's not just in physical appearance that owner and dog can correspond: sometimes, psychological choices that an owner makes – as seen here with clothing and collar – ensure even greater accord.

It's been suggested that it's possible to determine someone's personality by their choice of dog. What do you think *your* dog says about you ...?

- hairiness
- size
- sharpness of features
- attractiveness
- perceived friendliness
- perceived energy levels

– but no one single characteristic was obvious: choice of dog may have been influenced by what potential owners already thought about breed temperament differences.

In addition to physical comparability, it can be interesting to spend some time reflecting on your personal preferences, characteristics, and personality.

What sort of person are you?
Do you like taking long walks in the countryside, or prefer shorter strolls round and about your local area ... are you

a party animal or happier quietly reading? The type of personality you have may well influence what dog you have, and how he behaves. Personality traits encompass a huge range, including extremes – shy/outgoing; anxiety/stoicism; optimistic/pessimistic – and everything in-between. If you prefer gentle activities and stability and steadiness in your life, a high-spirited dog who needs lots of exercise and attention might not be ideal.

Your personality will influence your choice of canine companion, but that does not mean you might not choose a dog who is quite different to you. As in human relationships and pairings, the adage 'opposites attract' may also be true of human-canine partnerships, but analysing the personality match/mismatch when choosing a dog is interesting.

In another study (in collaboration with the Kennel Club of Great Britain and OnePoll) dealing with what your choice of dog says about you, an interesting variation between breed groups and owner personality features was noted. The study split the breeds into seven groups –

- Gundogs (eg Golden Retriever)
- Hound dogs (eg Greyhound)
- Pastoral (eg German Shepherd)
- Terrier (eg Staffordshire Bull Terrier)
- Toy (eg Chihuahua)
- Utility (eg Bulldog)
- Working (eg Doberman)

The study – by Dr Lance Workman and Jo Fearman of Bath Spa University – found that owners of dogs from the pastoral and utility group were more extroverted; owners of gundogs and toy dogs more agreeable; owners of utility, toy and gundogs more conscientious; owners of hound dogs more emotionally stable, and owners of toy dogs more open to new experiences.

Dr Workman said: "This study indicates that we might be able to make predictions about someone's personality based on the breed of dog that they choose to own. It seems that personality types are drawn to certain breeds."

He also suggested that: "The differences in personality factors found between owners of different breeds might arguably be related to the lifestyle of the owner. For example, more extroverted individuals might be better suited to the pastoral breeds such as German Shepherd or Border Collie, whereas those who are particularly emotionally stable might be suited to ownership of hound dogs such as a Beagle or Greyhound."

Your dog and you: lifestyle and psychology

What sort of lifestyle do you and your dog lead?

Consider and answer the following questions, making a note of your answers so that you can look back at them. When we take time to reflect on our lifestyles, valuable insight into ourselves and our relationship with our dog can sometimes result –

- how would you describe your relationship with your dog?
- are you both calm and settled, with a daily routine?
- are you a working owner who spends most of their time outside the home?
- how much time a day do you spend with your dog?
- who feeds your dog?
- who takes your dog for walks?
- how often do you take your dog out each day?
- what time(s) of day do you take your dog out?
- where do you go with your dog?
- do you take your dog in the car?
- does your dog go on holiday with you?
- is it just you and your dog in the family or all there others – possibly children – too?

Both you and your dog will change and mature over time, and a reassessment can be helpful to check that all is as good as it can be.

Reduce negative emotions in your life to benefit your dog and you
Sometimes when we are struggling and going through
a tough time in our lives, we might seek help to get our
lives back in balance. A visit to the doctor may result in the
suggestion of psychological counselling or therapy to try
and calm any unsettling emotions.

Some types of therapy teach practical strategies and
coping mechanisms to help get us back on track, and one
such therapy that has developed and, since 2006, become
particularly popular in the UK after publication of a major
research project covering wellbeing, is Cognitive Behavioural
Therapy (CBT).

What is CBT ?
CBT is made up of three parts –
C = Cognitive (our thinking and thought processes)
B = Behaviour (what we do)
T = Therapy (the change we make)

Our dogs watch our behaviour – what we do – and,
based on past experience, decide what is happening. For
example, packing a bag, putting on a coat, picking up keys,
usually mean we are leaving the house.

Our dogs don't *know* what are thoughts are when
this is going on – how could they? – but they make
deductions based on our behaviour and whether or not they
have seen it before.

Our dogs will interpret our behaviour and make assumptions based on previous
events. Alterations in usual patterns of behaviour or negative emotional responses
could affect how he behaves around us.

But what if, having got to this point, we decide that we don't feel well, and won't go into work after all? For a start, our behaviour will be different; we'll probably take off our coat, put down the keys and bag, and go back to bed!

We may be thinking "Oh no. I am aching and have a hot head. How will I be able to cope at work? I don't think I'll go."

CBT recognises that our behaviour and emotions are a result of our thinking.

Displaying different emotions and different ways of expression to the same dog can illicit very different reactions, even if he is usually a very calm dog, how he behaves will depend to a degree on our emotional state.

If our behaviour shows that we are experiencing negative emotions, such as anxiety, fear, sadness, or anger, our dog will pick up on that. This can affect their sense of security, and, in turn, their behaviour.

So, how can an understanding of CBT allow us to provide more settled lives for our dogs?

A change in behaviour in our dogs may be the first indicator that we need to stop and evaluate our emotional state. Has our dog noticed changes in our behaviour that indicate negative thinking, before we have? If we learn about CBT, how it works, how to use it effectively, and how it can help us recognise when we are experiencing negative or detrimental thought processes, then we can learn how to alter our thinking to prevent emotional upset.

Learning about CBT, and how it can help you be happier will be an investment not just for you, but your dog, too!

How do you feel?

Make no mistake: your dog will have a pretty good idea of how you are feeling. He is constantly scanning and assessing his environment for clues about what's going on, and if you are his main care-giver, he will watch you the closest.

Most of the time, we are not consciously aware of how we are feeling, because to constantly monitor ourselves would interfere with our daily lives. We are always thinking, however: there's an internal dialogue in our heads that we even sometimes verbalise by talking to ourselves. We are considering options and making choices all the time, and our dogs are aware of all the choices we make as they observe our behaviour, which is a direct result of our thoughts and feelings.

When I train counsellors, one of the first homework assignments I ask them to do is to keep a 'feelings diary' for a week; noting down some of the feelings that come and go. Some students accomplish this by setting an alarm to remind them to try and be aware – in the moment – a few times a day, so that they are conscious of what they are feeling. Others take time at the end of the day to reflect, and jot down remembered feelings.

Mainly students struggle with being able to 'name that feeling;' all of them of the opinion that somethig that sounds very simple is actually very difficult to do!

Try this exercise. In a notebook or on your computer, note the days of the week, Monday to Sunday, leaving enough space to describe your feelings for each day.

If you find it hard to 'name that feeling,' the following suggestions may help (you can copy this section so that it can be used more than once, if required) –

◯	calm	◯	apprehensive
◯	fearful	◯	contented
◯	happy	◯	annoyed
◯	worried	◯	anxious
◯	relaxed	◯	tired
◯	scared	◯	calm
◯	excited	◯	depressed
◯	sad	◯	hopeful
◯	appreciative	◯	irritated
◯	frustrated	◯	loving
◯	guilty	◯	strong
◯	insecure	◯	angry

We all experience many different feelings and emotions every day, some of which we're not even aware of on a conscious level: they simply come and go as we go about our business.

Sometimes, though, a prolonged period of negative feelings may begin to impact on us.

The way we think affects what goes on in our bodies: it has a physiological effect. Sudden feelings of anxiety cause our body to release into the bloodstream greater amounts of hormones such as adrenaline, for example, readying the body for a flight or fight response (as with all animals). This reaction is unsettling and detrimental to health ultimately, as our body is on high alert, causing stress and emotional imbalance.

A little knowledge of Cognitive Behavioural Therapy can help reduce some of these negative feelings. Consider these points –

- how we think is the result of how we view ourselves, the world, and our future
- we may think negatively at times, interpreting threats in our world
- when we feel threatened, our body reacts
- chemical changes in our body will affect our behaviour ...

The more we can understand exactly how we are feeling and why, the easier it will be to understand how feeling this way could impact on our ever-observant dog.

- ... which may perpetuate our negative thinking, resulting in a vicious circle

Now think about these –

- we *can* change the way we think about ourselves; the world; the future
- if we change the way we think, we can change the way we feel, and the way we behave

46

If we allow negative feelings to overcome us when we are around our canine companion, he, too, may develop the same negative outlook. Using CBT can help combat these feelings.

- CBT can help us work out what we are *really* feeling
- CBT can help us work out what the triggers are for our upsetting feelings
- CBT can help us understand how our thinking works
- CBT allows us to realise we have choices about *how* we think, and provides practical strategies to reduce emotional upset and distress

The Think ABC kit
I devised the following tool kit to help users calm unsettling feelings.
Step 1 – name that feeling: call this C (the emotional

consequence). The following suggestions may help (you can copy this section so that it can be used more than once, if required) –

- anger
- anxiety
- feeling low
- guilty
- frustrated
- criticised
- embarrassed
- disrespected
- resentful
- mocked
- worthless
- controlled
- pressured
- alone
- ignored
- rejected
- unloved
- blamed
- cheated
- scared
- discouraged
- abandoned

Step 2 – identify the situation that gave rise to the feelings: call this A.

In-between A and C, of course, must be B –

Step 3 – work out what you were thinking at the time the situation occurred: call this B (your beliefs/thinking)

How we think can be a reflection of our beliefs about how things should be in the world; how other people should behave; how we should be treated, and our views on life in general.

In CBT, we say that a situation or another person are not the reasons why we sometimes feel bad, but rather what we *think* about the situation or the person is what gives rise to our negative feelings. In that case, A does not cause C, but what and how we think about the situation (B) decides how we *feel* about it (C).

Often, we have quite strong beliefs about how we should be treated by other people and life; what should happen to make life 'fair,' or how we should behave in order to be a 'good' person. When life doesn't live up to our expectations, we can become upset, giving way to unsettling feelings that "It shouldn't be like this!" But if it's not possible to change the situation or event that has made us feel this way, continuing with this belief will only result in further distress and emotional imbalance.

And all the while, don't forget, our dog is watching us

Just as people feel empathy and can share the pain that another is feeling, it is thought that dogs have this ability with their owners. The notion that our dog struggles when we do is a powerful one ... (and opposite).

struggle with the difficult situation or interaction with others, and our upset becomes *his* upset!

When faced with this situation, one way we can help ourselves is to stop and consider how we can change our thinking, and give ourselves time to rationalise our thinking.

Changing our way of thinking from "It shouldn't be like this!" to "I don't like what is happening, and I would prefer

48

that things weren't this difficult at the moment, but how is becoming upset helping me? Or, in fact, my dog?"

By changing B (beliefs and/or thinking), the emotional consequence (C) will also change to something less upsetting. For example, instead of anger, we may feel annoyance, which will not impact quite as dramatically on our behaviour, reducing the negative vibes that our dog interprets. In addition, a less heated response will allow us to consider the matter more rationally, with a resultant appropriate response

The foregoing is a simplified introduction to Cognitive Behavioural Therapy, but should be sufficient to raise awareness of how your emotional state affects your dog, and, in turn, his emotions and behaviour.

For a more detailed explanation and help with applying the Think Kit, please see my book *Introducing Cognitive Behavioural Therapy (CBT) for Work: A Practical Guide*, and other books on the same subject.

Visit Hubble and Hattie on the web: www.hubbleandhattie.com & www.hubbleandhattieblogspot.co.uk • Details of all books • Special offers • Newsletter • New book news

50

Psychological development and growing up experiences

Previous chapters have explained how being aware of how we, as owners, impact on our dogs can be very useful. Our feelings and subsequent behaviour can have a huge influence on our dog's wellbeing and sense of security.

Perhaps we should invent a new therapy for our dogs called Canine Behavioural Therapy, or CBT for Dogs, which would encourage a two-way link between us: a true meeting of minds.

How might this influence a dog as he grows and develops?

In this chapter we'll look at the stages in a dog's life, both physical and psychological. Our influence on how our dogs develop is constant and profound; right from birth, a dog has to interact with people.

When our dog comes to live with us, he already has some history, depending on his age. Having an understanding of how this colours his temperament and behaviour, and how well or otherwise he settles into our family, will help us to help him become a happy and well-adjusted part of that family.

The psychology of human/canine development

Many factors influence the way we grow up, and early experiences, in particular, are known to have a big impact on our developmental areas –

- physical: body development, genetic influences
- emotional: psychological development
- intellectual: thinking/mental processes
- social: interaction with others and the environment

Our dogs derive great benefit from the security they feel when in the presence of a loving owner.

– and it is the case with dogs, too –

- physical: body development, genetic influences
- emotional: changes in blood chemistry that create sensations, fear, aggression, anxiety, etc
- intellectual: thinking/mental processes
- social: interaction with other dogs, people and the environment

CBT could encourage a meeting of human and canine minds, allowing for a closer connection between them.

Stages in a dog's life

The birth history and development of a dog can greatly influence how he develops.

The ideal scenario would be to have a complete knowledge of a dog's background, but often this is not possible, depending on where our dogs are acquired from.

In response to the ever-increasing number of dogs who find themselves without a permanent home, many rescue centres exist, sadly, and in the UK the two main ones

If choosing to take an older dog into your home, it's important to remember that he will already have encountered people and made judgements about the world. We need to first understand and then adapt to possible challenges this may cause.

A dog's initial interaction with the world can shape his personality and how his behaviour develops.

are Dog's Trust (founded in 1891 and formerly called The Canine Defence League, the Trust has eighteen rehoming centres, and usually around 18,000 dogs in its care every year), and Battersea Dogs and Cats Home, as well as many regional centres.

Dog rescue centres are to be found in many other countries, too, and huge efforts are made to raise funds to help save dogs from starvation and homelessness. I have taken holidays where big travel companies include in their list of excursions an opportunity to visit a dog sanctuary, and take some of its residents for walks, in an effort to make people aware of the centres, and help if they are able to.

We seem to have an inbuilt compulsion to want to save and care for those animals who have been neglected or badly treated.

So, what might it mean, to give a rescue dog a home?

Not surprisingly, a rescue dog's psychology may well have been affected by past experience. Although his history may not be known, what has happened to him during development will have influenced and shaped his current state of mind. Rescue centre organisations are expert at knowing what these animals need, and provide the best possible care for all their needs.

They will talk to potential new owners; consider all aspects of their situation, and what they can offer a dog, plus advise on a dog's suitability for rehoming. Matching every dog as carefully and precisely as possible to potential rehomers, this level of professionalism gives both dog and new owner the best chance of a harmonious and positive outcome.

Of course, rescue centres cannot know all of the past circumstances of every dog, but the behaviours and reactions that each exhibits – which are monitored after a dog is taken in – provide good clues to how they are likely to behave in different scenarios, and what the best setting is for each.

For example, a dog who has been left on his own a lot might become very anxious if he goes to a home where the house is empty a great deal. Much more suitable for a dog like this would be a home where at least one person is there most of the time, so that the dog can relax and feel secure.

After a dog is born, what are the important and influential stages in his life?

The first few weeks of a puppy's life are vital for imprinting and bonding with his mother, learning to feed successfully, discipline, socialising with his littermates and with people, and forming attachments.

It is important that puppies remain with their dam for at least eight weeks after birth, whilst they go through critical development periods –

Two-four weeks (transitional period)

As a puppy's senses come into play in his second week of life, he will begin to move around, exploring his immediate environment.

He will begin to really interact with his mother and siblings, learning what it is to be a dog. Mum is instrumental in teaching discipline and boundaries: the different personalities of the pups begin to emerge, and as they tussle for position and dominance in the litter, their mother soon lets them know who is boss.

One train of thought is that if the mother is not around to discipline the more unruly pups, this could lead to problems when the dog becomes part of a human family.

Four-twelve weeks (socialisation period)

Puppies enter the socialisation period at the end of the third week of life, and this continues until around week ten. This is an important time for the puppies as so much learning occurs, and what they experience and learn now can hold them in good stead for the rest of their lives.

The most critical period – six to eight weeks – is when puppies most easily learn to accept others as part of their

Giving a rescue dog a second chance can be infinitely rewarding, but we should give consideration to how to behave around these animals, and ensure they are placed in the best environment for them to thrive.

family, so it's important that they are acclimatised to gentle handling, when checking for health issues, maybe, and to establish their trust in us. It has been found that when this process occurs naturally, and with care and consideration, this paves the way for a dog to develop in a secure and balanced way.

Weaning, typically, is complete by week eight.

Eight-twelve weeks

Puppies often go through a 'fear period' during this time. Instead of meeting new or familiar people and objects with curiosity, they react with fearfulness. Anything that frightens them at this age may have a lasting impact so take care that there are no negative experiences, such as lack of parent interaction, food deprivation or limited opportunities to play and interact with littermates.

From eight weeks of age a puppy can leave his canine family. Interaction with siblings and his dam will have helped him learn bite-inhibition and his place in society, and this, together with positive contact with people, mean he has a very good chance of developing into a secure and well-adjusted dog.

Genetic influence

Earlier in the book we looked at how genetics affected the canine temperament. Some breeds are more suited to certain roles than others: they may be instinctively territorial and protective, so make good guard dogs, for example, whilst others with a softer temperament make ideal companions. Canines with natural abilities such as herding are happiest when working.

These genetic traits are inherent, but to what extent they develop depends on a dog's early environment and training.

Can we test puppies for early signs of these atttributes? Yes, we can: tests have been available since the 1960s.

The tests were developed by Clarence Pfaffenberger, originally to gauge the suitability or otherwise of dogs as potential Guide Dogs for the Blind. Pfaffenberger tested puppies at the eight-week stage, and through selective breeding programmes, careful observation of the development of the puppies and attention to their needs, the success rate for choosing puppies who became good guide dogs rose from 9% to 90%.

Training Guide Dogs is a lengthy and costly exercise, and it was important to be able to spot a dog's potential as soon and as accurately as possible, and Pfaffenberger discovered there was a critical window in which to place the puppy in his new home after testing. Of the puppies who were placed in their new home in the first week, 90% went on to be successful guide dogs. Those who were not placed until one or two weeks later were not quite as successful and for every week the animals remained in kennels after testing the success rate dropped, until after three extra weeks, only 30% of the animals there went on to become Guide Dogs.

This new information was very interesting, and more importantly, brought the realisation that a dog's early life, and when and how he is placed in his new home, has a significant bearing on future development and behaviour.

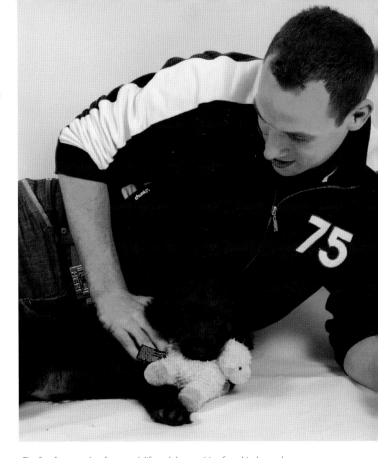

The first few months of a puppy's life and the transition from his dam to human carer are instrumental to his development.

Some breeders have their puppies tested at this early three-month stage stage to gauge potential.

In 1975, animal behaviourist William E Campbell adapted the tests to assess a dog's suitability as a family dog, based on temperament. Campbell's findings are as follows, and are typical of their time. Opinions and methods have changed a great deal since then, most notably in connection with the old, disproven dominance theories and myths. *In no way do the author or publisher subscribe to these outdated, often cruel ways of interacting with a dog.*

Excitability v Inhibitability

In this instance, an 'excitable' dog is taken to mean one who is very responsive to external stimuli. Field trial retriever dogs have this trait, for example, and it means they are very

Opposite & above: Some dogs can be coaxed and guided with toys because of their inherent 'excitability' trait.

sensitive to and aware of what is happening all the time, posied to retrieve their prey when it lands.

Dogs with inhibitability traits were more likely to exhibit better self-control, responding only when given an appropriate cue to do so. Campbell cites the German Shepherd as an example, here.

Active v passive defence reflexes

Regarded as an inherited trait of reacting to stress by running away, freezing or biting. If your dog displays active defense reflexes, special training will be necessary in order to reprogramme these natural tendencies.

Dominant v submissive

If a dog has a tendency to dominate, some of the behaviours he may show that would seem to indicate this are –

- direct eye contact
- walking with head or hackles up
- biting behaviours, growling
- mounting

Direct eye contact and a confident nature suggest that this dog would have led her social group, had she remained with her littermates (note her beautiful, differently-coloured eyes: a condition known as heterochromia).

Your dog and you – understanding the canine psyche

The submissive dog usually recognises a leader and is accepting of this. Behaviours may include –

- rolling over
- crouching low
- avoiding eye contact
- tail tucked between legs
- avoiding confrontation
- willingness to obey commands

Patience, understanding and a gentle touch can help build this dog's confidence, and habituate him to living in a busy environment.

Independence v social attraction

As mentioned previously, the friendliest, most empathetic dogs are the ones who, in general, have evolved to live with people. Some dogs are naturally very independent, and not interested in human company, though it is important to be aware that a dog who has had an unhappy early life, or who has experienced negative interactions with people, may also show this tendency.

On the other hand, the socially-attracted dog likes to be in human company. He may follow the family around, loves to play, be held and cuddled, and usually at ease and relaxed. Some dogs are especially bred for these characteristics as they make such good companions for us .

Other dogs show completely contrasting behaviours, suggesting a more submissive nature.

Sensitivities

Two other inherited characteristics have been identified by researchers and authors Elliot Humphrey and Lucien Warner.

Sound

When there is a loud noise or a sharp report, sound-sensitive dogs may respond by crouching, running away, showing excessive fear, or urinating. Fireworks, gunshots, a bird scarer or shouted commands will impact detrimentally on these dogs.

Touch

Dogs can respond differently to being touched. Some react hardly at all, whereas others are very sensitive. A touch-sensitive dog behave defensively, exhibiting biting or freezing behaviours, or running away.

There are many different things to take into account when considering our dog's psychology as he has developed and grown. As we are coming to appreciate more and more, dogs are very special animals who, over many years, have evolved to be our companions. They do experience emotions, possess the ability to think and reason, and suffer emotionally and physically, just as we do.

We dictate almost every aspect of our dogs' lives: when they can eat, go to the toilet, play and exercise. In the past, we've thought that relaxing that control would result in our dogs challenging our authority, but it's usually the case that they only want to be what they always have been: our workmates, companions, and much-loved family members.

We owe it to them to really pay attention to what it is our dogs are trying to tell us, and our relationship will grow and blossom as a result.

Dogs can develop sensitivity to sound or touch, which needs to be handled gently and carefully to maintain the harmony between the two of you.

How do we know when our dogs are stressed or in distress?

If canine behaviour is sometimes a result of interaction with the environment, what signals are being picked up on to make dogs behave in a distressed way?

Is it possible that some of the following unpleasant emotions are responsible –

- anxiety
- anger
- fear
- frustration
- pain
- depression
- low self-esteem – perception of pecking order in the social group

Canine stress/distress signs and symptoms

Dogs are very similar to people in that the ways they show stress can vary. The make-up and personality of an individual dog will influence behaviour which reflects the fact that he is finding his environment stressful. The basic survival mechanism – 'fight or flight' when faced with a threat – is inherent in both dogs and people. Biochemical changes in the dog's body – increasing levels of adrenalin cortisol – determines behaviour to combat the perceived threat.

- Fight behaviour: lunging forward; barking; attempting to bite
- Flight behaviour: cowering, fleeing, hiding

Interestingly, fight or flight are not the only options a dog has when facing a threat: he may choose to freeze (as if invisible); display displacement behaviour (trying to defuse a stressful situation by engaging in a totally different and unconnected behaviour, such as rolling on the floor or attending to an itch), or 'faint' (not actual fainting but when a dog drops to the ground and refuses to budge).

Reactions to stress can vary: some dogs will direct their stress at people or their possessions; others may internalise their stress and eventually may become ill, and we don't always recognise the signs of this happening.

Typical stress reactions can include –

- barking and howling
- destructive behaviours – chewing, biting, tearing up things, digging (may be directed at inanimate objects or at people)
- pinned ears
- panting
- soiling in the house
- growling, snarling and nipping
- stress yawns
- lip and nose licking
- becoming unwell – digestive upsets, skin problems, gnawing/excessive licking of limbs
- excessive scratching
- excessive sniffing

Other, more subtle behaviours and body language include –

- 'shaking off' – scrolling the fur, almost as an attempt to shake off an uneasy situation, such as a visit to the vet
- low tail – can indicate fearfulness

Dogs can demonstrate stress in different ways: in the form of aggression and teeth-baring, or by withdrawing from the situation.

When confronted with doubtful situations, dogs will deal with these with differing levels of efficiency. It can even cause them to become ill, just as with us.

- low body posture
- changes in eating habits
- pacing, restlessness
- signs of strain around the eyes and mouth; furrowed brow
- shedding hair more than usual
- avoiding people – looking/turning away
- sweating from paws – indicated by sweatmarks on the ground or floor
- dilated pupils
- slow, deliberate movements: cautious and watchful
- whale eye (when the white of the eyes show)

Something to watch out for is if a stressed dog suddenly stops panting and closes his mouth: this could be a sign that he is about to nip or bite.

Calming signals

A dog uses calming signals in an effort to prevent unpleasant things from happening – avoiding conflicts, avoiding threats, calming people and other dogs, dispelling nervousness, fear and anxiety, and also saying "I am no threat." All dogs use these signals as a universal language, which mean the same whether the dog is an Akita, a German Shepherd or a Neapolitan Mastiff.

Typical calming signals –

- head turning – dogs dislike head-on approaches from other dogs, animals, and people, preferring

An intense stare or sudden closing of the mouth can indicate stress – and impending aggression.

to meet in a curve. Turning the head avoids a direct gaze, which can be perceived as threatening. Have you noticed that when you take pictures of your dog he turns his head away?

- softening the eyes – again, a direct gaze is a sign of threat so softening the eyes by squinting is good doggy manners.
- turning away – often done to avoid a threatening situation.
- nose-licking – how many of you have pictures of dogs licking their lips? Very likely the flash of the camera, or the camera itself made them uneasy, and they licked their lips in a calming signal.
- freezing – feeling intimidated by another person, dog or situation, a dog may freeze in an effort to show he is not a threat, especially if a bigger dog is around.
- walking slowly – quite often seen when a dog owner scolds his dog for not coming to him. The dog does obey, but at a very slow pace, which makes the dog owner more frustrated!
- play bow – often an invitation to play to demonstrate the casual nature of the action. Frequently used to make friends with a shy or suspicious dog.
- yawning – if your dog yawns, it's quite likely that he's not tired but giving you a calming signal.
- sneezing – probably not a cold but a way of diffusing a worrisome situation.
- sniffing – out of context sniffing is a calming signal – a displacement activity.

Tension builds in dogs just as it does in us, and we should be vigilant in spotting signs of it in our own companions, in the interests of safety and our dog's wellbeing. Stress can creep up without us realising it, and we may 'snap' and become irritable, angry, upset, fearful, and even sick. Our dogs are watching us all the time: if we are stressed, they pick up on this and can assimilate the nervous tension, becoming anxious and stressed themselves as a result.

We are totally responsible for the environment in which our dogs live, and identifying the source of our dog's unease is therefore essential.

Survival dependency

Dogs depend on us for their survival, and any change in routine, environment, relationships and behaviours can give rise to a feeling of insecurity.

Dogs are masters at body language, and show us in so many ways – as described – when all is not well. By being aware of the various signs and signals dogs use to convey how they're feeling, we can make the connection between exhibited behaviour and what it is that's causing the anxiety.

Aggression in dogs

Interesting research into aggression in dogs has been carried out by the University of Bristol School of Veterinary Science, which released its findings in January 2014.

The first question to consider is: do we consider that aggression in dogs is a sign of distress? Many would argue that aggression is a natural state for dogs in connection with territory control, survival in the hunt for food, and procreation of their species. Given that we take care of these considersations for our domestic dogs, is aggression a necessary or acceptable emotional state for dogs in society?

It is a fact that dogs have the potential to harm and possibly even kill, so it's important to understand canine aggression: whether some breeds are more aggressive than others; who or what the aggression is usually directed at; when it might occur, and how to prevent it happening. In the aforementioned study, 4000 completed questionnaires revealed that –

- aggression towards unfamiliar people was reported more often than aggression towards family members
- approximately seven per cent of owners reported that their dog had barked, lunged, growled at or actually bitten visitors to the house
- five per cent reported that these behaviours occurred when out on walks
- three per cent of dogs nehaved aggressively toward family members

An important finding was that showing aggression seemed to be situation-specific: this occurred in one situation; not generally. This seemed to indicate that, as

continued page 66

Our canine companions rely on us — as leader of their social group — to provide an emotional example for them to follow.

Understanding our dog's emotions, in tandem with our own, can result in a much happier and mutually-fulfilling partnership.

this aggression was obviously a learned response, the dogs should not necessarily be labelled vicious.

Other findings from revealed that dogs who went to puppy training classes were –

- 1.5 times less likely to be aggressive with strangers
- were better socialised at an earlier age
- were twice as likely to be aggressive if owned by the under-25s, rather than the over-40s
- were better behaved with an older, more mature owner
- were twice as likely to be aggressive to strangers, and three times as likely to be aggressive toward their owner if trained using the punishment-based, negative-reinforcement method and shouted at

It does seem that aggression can stem from fear and anxiety. A dog who is subjected to a harsh and unkind training regime will respond with fear and anxiety. One who is handled kindly and compassionately, with reward-based, positive-reinforcement training, should be a happy and well behaved animal, regardless of breed.

The study concluded that whether or not a dog is neutered has no bearing on aggression levels.

One of those involved in the study – Doctor of Educational Psychology Anthony Carboni – concludes it's not possible to state that a Pitbull is more likely to be aggressive than, say, a Poodle, because breed type is just one small factor. Other factors are training; treatment by owner; environment; owner's knowledge and experience.

The more we can learn and understand about our dog's psychology, and use what we have learnt to maximise his sense of security through well thought out training, care and consistency, the more stable, predictable and content our dogs will be.

When we encounter canine behaviours that we don't understand, and which may be undesirable to us, and/or distressing for our dogs, it is important that we are willing and open to the idea of asking for help and advice.

Our dogs and us – a dynamic, interactive partnership: the better we understand, the happier we will be.

In the next chapter we look at ways to recognise when WE are feeling stressed, and provide some practical tips on how to reduce this.

We have learned what some of the signs of distress in dogs are, and how to recognise them, so now we're going to do the same for ourselves. Armed with this valuable self-knowledge, we can also think about how this may affect our dog.

Often, we are so busy in our daily lives that, although aware that things aren't quite 'right' within ourselves, we don't have the time – or maybe even the motivation – to examine this further.

A dog can be man's best friend, but first we have to be our own best friend – and we owe it to our dog to be the best we can.

Remember – If we're not happy, our dog may well not be happy.

Aggression was one of the negative signals that dogs living with people displayed, and one of the signs of unsettled emotions in dogs is aggression, which can be caused by –

Understanding ourselves, and knowing why we do the things we do, is the first step to really being able to truly connect with our canine friend.

- environment
- genetic history
- training
- treatment and handling
- owner situation

All of the above factors influence how a dog is feeling.

Let's look at how similar circumstances to these impact on us. The most common distressing emotions which those people I have treated in therapy complain of are –

- anxiety
- frustration
- anger
- guilt
- depression
- low self-esteem – perception of status with family, friends, work, society; the world at large

Sound familiar? There are similarities in the above list to that we listed for our dogs in the previous chapter.

The human animal has the same basic ancient reaction to stressful situations as dogs – fight or flight – and our bodies respond to an alert by producing large quantities of certain chemicals –

- adrenalin
- noradrenalin
- cortisol

Adrenalin and noradrenaline prepare the body to react to danger, priming the muscles to provide extra strength. Certain visible physiological changes occur, which include pupil dilation, hair on the body standing on end (piloerection), muscle tenseness (including teeth clenching), and dilation of blood vessels in the face and neck so that these areas become red and scary-looking.

Cortisol is an important and helpful part of our body's response to stress, providing a quick burst of energy, increased immunity, heightened memory function, reduced sensitivity to pain, and maintenance of homeostasis (metabolic equilibrium) in the body. It's important that the

Reactions to distressing emotional states can be very similar in ourselves and our dogs, with anger and anxiety common emotions in both species (and opposite, left).

The 'fight or flight' response is also shared between human and canine. The perception of our actions in these situations can lead others to misunderstand our true personalities and intentions (and overleaf).

body's relaxation response returns functions to normal as quickly as possible, however, as prolonged levels of cortisol can give rise to a state of chronic stress, resulting in lowered immunity, impaired cognitive function, blood sugar imbalances, higher blood pressure, and suppressed thyroid function. Depressed spirits are often associated with high levels of cortisol, which can also be present in the body in greater amounts following a bereavement, loss of job, or change in relationship status.

Do you recall the changes that take place in a dog's body during a potentially threatening situation or continued stressful state? Can you see the similarities?

Signs and symptoms of stress in people
These can be many and varied, but are common across the human species –

- racing heart
- increase in blood pressure
- sweating
- breathlessness
- 'butterflies'
- urgent need for the toilet
- light-headedness
- trembling/shaking

- muscle/joint weakness
- exhaustion
- changes in diet
- insomnia/sleep disturbance
- reliance on alcohol or other 'prop'
- blurred vision
- increased suscepibility to illness
- immune system conditions: arthritis
- digestive problems: IBS
- headaches
- low spirit/depression
- irritability
- anger
- increased sensitivity
- reduced frustration level
- restlessness/need for physical release
- loss of sense of humour
- making mistakes
- loss of concentration
- impulsivity
- overreacting to others
- negative outlook
- jumpiness
- shouting at others
- defensive and self-protective
- self-blame
- snapping at family and work colleagues
- lethargy
- lack of interest
- emotional outbursts
- hair chewing/nail biting
- frowning/scowling
- shunning company/isolation

Think back: can you identify any similarities between our symptoms and behaviours and those that dogs exhibit? Yes? Quite right! The table overleaf looks at some of these.

Behaviours that we sometimes try to train out of our dogs can actually be natural responses to their surroundings and situations, and we should be careful not to stifle their natural instincts.

Dogs	People
Barking/howling	Emotional outbursts
Chewing/biting	Irritability/anger/restlessness
Panting	Breathlessness
Soiling	Urgent need for the toilet
Growling/snarling/nipping	Irritability/snapping/shouting
Stress yawns	Stress yawns
Lip and nose licking	Hair chewing/nail biting
Digestive upsets	Digestive upsets
Excessive scratching	Jumpiness; inability to keep still
Shaking off	Shrugging/apparent nonchalance
Low tail/body posture	Low spirit/depression/slumped posture
Changes in eating habits	Changes in eating habits
Pacing	Restlessness/need for physical release
Signs of strain around eyes and mouth	Frowning/scowling
Increased hair shedding	Hair loss (alopecia)
Avoiding people	Shunning company/seeking isolation
Sweating from paws	Sweating
Dilated pupils	Dilated pupils
Slow, deliberate movements	Defensive and self-protective
Excessive sniffing	Increased sensitivity
Pinned ears	*Hmmm, perhaps not!*

The foregoing comparison clearly demonstrates the similar reactions we experience when the heat is on.

When we notice these behaviours in our dogs, we often decide that they are undesirable, and wonder how we can change the behaviour. But in light of what we've just discovered, it would seem that these behaviours are actually very natural – and quite normal – for our canine friends.

Remember: our dogs are watching us all the time; looking for clues and signs of what's expected of them. After all, their very survival depends on living with us as harmoniously as possible, so that we continue to be their care-givers.

Likewise, children depend on others – usually their parents – for their survival, as least until they are old enough to take care of themselves.

As well as the practical aspects of this, such as food, warmth, and shelter, the emotional development of children is very important, and babies frequently exhibit 'separation anxiety' when their mother is absent.

If we can harness the power of controlling our own emotions and how they reflect on our actions, we are much more likely to have a happy and well-adjusted dog with an even temperament.

This separation anxiety has been recognised in dogs, too, maybe because, knowing that they are dependent on us, they become anxious if they feel their security is threatened by our absence.

Of course, as children grow into adults they are able to reason and rationalise, taking responsibility for their security themselves: something that dogs are unable to do

The friendliest, most versatile breeds survived because they learned to co-operate and live alongside us. Creatures of habit, dogs like to have boundaries, routine, and consistency, and when there is harmony and empathy they feel secure and settled.

If this harmonious state is threatened, the resultant feeling of insecurity and maybe even fear causes a reaction; most usually a change in behaviour. If we can be aware of as many factors as possible that may cause our dogs to react – possibly negatively – including our own feelings of stress, insecurity and negativity, we will go a long way toward truly understanding and empathising with this most unique of animals.

This chapter has highlighted some of the symptoms and behaviours that we may be experiencing which our dogs could be picking up on. If we first begin to recognise and accept that these are due to unsettling feelings within ourselves, we can hopefully rectify the situation, showing our dogs that all is well, and they have no reason to worry or feel under threat.

Visit Hubble and Hattie on the web: www.hubbleandhattie.com & www.hubbleandhattieblogspot.co.uk • Details of all books • Special offers • Newsletter • New book news

74

We are beginning to understand and appreciate much more how owning a dog – or being around dogs and other animals – benefits our health and mental well-being; even to the extent that life expectancy is extended.

A dog can add structure, purpose and meaning to our lives, providing a reason to get up and get going in the morning; take health-giving, envigorating walks; interact with other dog owners; care for and take responsibility for another living creature.

The benefits to our species

We are hardwired to benefit from positive feedback: caring for others, initiating 'random acts of kindness,' and doing good deeds, all stimulate our brain's centre of emotions, releasing the neurotransmitter serotonin, which, amongst other things, can influence our mood and social behaviour. Because of the fact that we benefit ourselves from such acts, it is believed that altruism does not exist. In terms of survival of our species, this makes sense: we are more likely to succeed and prosper if we live co-operatively.

How individuals benefit

Studies and research concerning how living with dogs benefits people – both in society and as individuals – have revealed some interesting facts.

It is known that those of us suffering from depression or low spirits are more likely to have lower levels of serotonin, although scientists are unsure whether decreased levels of serotonin contribute to depression, or depression causes a decrease in serotonin levels. Sometimes, medication can beneficially affect serotonin levels, and counselling and taking exercise can also help (research has suggested that exercise can increase brain serotonin

The benefits of owning a dog are irrefutable. Dogs have the ability to add structure and enrichment to our lives.

function). From the foregoing, we can see how living with a dog might contribute to our wellbeing.

Living with and caring for a dog helps us in so many ways, such as –

Touch

Humans have a basic need to be touched. When we touch

Owning a dog gives a reason to exercise – and release the 'feel-good hormone' serotonin. Of course, they also provide that all-important companionship.

http://medicaldetectiondogs.org.uk).
- a reduction in stress hormones
- a boost in the feel-good hormone, serotonin
- lowered blood pressure

Exercise
Taking exercise with our dogs can benefit us physically by –

- strengthening muscles and joints
- increasing fitness levels
- lifting mood
- promoting sleep
- promoting relaxation, particularly if we walk our dogs in natural surroundings

Illness
Dogs have proved especially beneficial in the detection of illnesses, such as –

- seizures – some dogs can be trained to recognise the changes that signal an imminent seizure (epileptic fit).
- food allergies – dogs can be trained to detect the smell of, for example, peanuts in the air, and alert the person of the danger.
- eczema – early findings of a study published in the *Journal of Pediatrics* revealed that children living with dogs were less inclined to suffer from eczema.
- psychiatric illness – a study involving 140 Doberman Pinschers discovered a gene for Canine Compulsive Disorder (CCD), similar to the gene present in humans for a condition such as autism. Reporting in *Medical News Today*, scientists said they had found that human and canine genetic pathways were similar for certain behavioural problems. The research aims to identify risk factors for certain diseases in dogs, which will aid understanding of canine health as well as its links to human health.
- cancer detection – in 1989 a case was reported of a dog who kept sniffing at a mole on his owner's leg, which was subsequently found to be cancerous. Not only that, dogs can also detect breast, ovarian, bladder, lung and colon cancer.

our dogs – stroking, patting, hugging, cuddling – the following effects have been noted –

- detection of low blood sugar – an article in the *British Medical Journal* in 2000 reported that dogs who lived with diabetics could detect low blood sugar before the diabetic did. The dogs would act differently, and even try to nudge their owner into eating. Researchers thought the dogs might be picking up on subtle changes in the owner's scent, or possible muscle trembles (see

The simple act of stroking our four-legged friend can help reduce stress and lower blood pressure.

Walking our dog can have lasting effects, providing the motivation to get out of the house and exercise ourselves.

A Labrador called Panda, who had been specially trained to detect colorectal cancer, correctly detected 33 out of 37 samples of colorectal cancer in a test.

- pet owners have been found to have lower triglyceride and cholesterol levels than those without (high levels can indicate heart disease).
- heart attack patients survive longer if they live with

pets, especially dogs. Linda Handlin and Professor Kerstin Uvnas-Moberg, of the Karolina Institute in Sweden, found that owning a dog could not only prolong life, but also reduced the likelihood of a heart attack, and increased survival rates after a heart attack by 75%.

The pair believes that human-animal interactions

Dogs of many different breeds have been found to have certain natural abilities, and have been subsequently trained to help us by detecting illness, and sniffing out drugs and explosives.

produce bio-chemical changes in us that are similar to those experienced in mother-baby interaction during breastfeeding.

Oxytocin, a powerful bonding hormone, acts as a neurotransmitter in the brain, and levels rise when we hug or kiss a loved one. It is also produced when a woman breastfeeds her baby, helping the pair to bond, and increasing the baby's chances of survival.

Handlin and Uvnas-Moberg had a hunch that a similar thing happens when we interact with our dogs. Taking blood samples from people before and after a petting session with their dogs, they tested the blood at one minute and three minute intervals, and could see the oxytocin levels rise and peak.

Apart from its bonding properties, oxytocin can reduce blood pressure, increase tolerance to pain, and reduce anxiety, claimed Handlin.

In another study entitled 'Dogs have feelings of love, too,' Handlin and Uvnas-Moberg found that not only did oxytocin levels rise in people during an hour-long stroking and cuddling session with their dogs, but so, too, did the dogs' levels. This rise was noted just three minutes into the petting session, and after fifteen minutes oxytocin levels of people and dogs had risen in a similar way. A further benefit was that cortisol levels reduced in those owners who felt their relationship with their dog was positive, as well as in their dog.

The other interesting fact about the findings of this study is that the effect of oxytocin on both people and dogs is that it tends to oppose the behaviours which facilitate the 'fight or flight' response.

Not only do we feel calmer and more relaxed, it seems, but so, too, do our dogs!

- studies showed that pet owners over 65 years of age make 30% fewer visits to the doctor than those who don't own pets. As dogs are in the majority for choice of pet, we can safely assume that they have a big influence on the statistics.

Psychological benefits
- social interaction with other dog walkers reduces feelings of isolation
- taking responsibility for another helps keep us focused.
- unconditional love – a sense of stability in our busy lives; dynamic relationships; uncertainties in life; jobs.
- reduced tension can help with depression; illness. It has been found that people who are depressed and in low spirits are more likely to have lower levels of serotonin. Sometimes medication is used, together with counselling and, perhaps, exercise, which is known to increase serotonin levels.

Quite evidently, living with a dog naturally provides us with opportunities to interact and benefit from their companionship.

Laughter and dogs
When we smile and laugh, our bodies release stress-busting hormones such as serotonin, whilst cortisol – which

Scientific study has shown that during a petting session, not only do we become calmer and less stressed, but so, too, does our dog ...

becomes unhelpful to us if it remains too long in our system – is reduced. In addition, our muscles receive a great work-out, leaving us feeling relaxed and happy.

For many of us, just seeing a dog raises a smile. We warm to dogs so easily: who isn't cheered by an enthusiastic greeting from their excited dog when they return home?

Our dog's enthusiasm and obvious delight when we arrive home – and their desire to be close to us – never ceases to raise our spirits. We feel wanted; we feel close to another living animal: a wonderful connection that's a special and precious bond.

Dogs have a natural ability to make us smile, and sometimes even laugh out loud with their natural zest for life and unconditional affection, dramatically affecting our mood and health in a very positive way.

Your dog and you – understanding the canine psyche

Living with dogs: benefits for children

Children love playing with dogs, enjoying a natural synergy to play, run around, tumble and jump, cuddle, hug, laugh and giggle.

Dogs never judge or say critical things, and are always pleased to see us, happy to loll about sharing relaxing time. For children they are the ideal companion, friend, and confidant.

Children who share their lives with a canine companion benefit both physiologically and psychologically. As mentioned previously, it appears they are less at risk of conracting asthma, eczema and allergies, and take less time off school through sickness (those children with pets had, on average, an extra nine days at school compared to those without).

It seems that living with an animal helps the body build a stronger immune system with which to fight off infection, according to the University of Warwick, which carried out studies of children who lived with pets: having a dog or cat exposed children to more infections in earlier years, which helped build a stronger immune system.

Stable immunoglobulin levels were more

Health benefits for children can be even more striking. Growing up with a dog around can strengthen their immune system, helping them to become healthier adults.

evident than in those children who didn't live with pets. Immunoglobulins (antibodies) are proteins that circulate in the blood, providng protection from disease by binding to foreign proteins, inhibiting activity and forming large complexes which are rapidly cleared from the circulation.

Babies who live in homes with dogs have fewer colds, fewer ear infections, and need fewer antibiotics in their first year of life than those raised in pet-free homes, Finnish researchers found, although the reasons why were not totally clear, though the suggestion is that immune systems mature best when infants are exposed to germs in just the right amounts.

Psychological benefits

As children develop they learn how to take responsibility for themselves and others. Growing up with a dog can help them develop a positive sense of self, as being a valued member of the family, helping to take care of the family dog. This can help give them a sense of security and well-being.

Dogs can help stimulate a child's brain through play and care, as well as provide a source of relaxation and calmness.

For children who may have difficulty communicating in the conventional way, having a non-speaking companion can facilitate their own communication by stimulating non-verbal communication such as hugging smiling, laughing, movement, and generally expressing what they are feeling with body language.

Living with dogs: benefits for the elderly

As we get older we may not be as strong and energetic as we once were, but owning a dog encourages us to get out for exercise every day. A well-trained dog can let off steam off the lead in suitable environments, whilst his owner takes a more moderate route around the park or field. Being outside gives a natural boost to the spirits whether in town or country, as well as opportunities to interact with others. Dog owners already have something in common, and can be a natural point of contact when out walking. This is especially important today, with extended families living miles apart, and friendship groups possibly reducing after we have retired, for example.

And the same applies to those who are out of work for whatever reason, as well as school- or uni-leavers who are

Undoubtedly highly intelligent, dogs can be trained to help people in need, enabling them to do things they would struggle to do on their own, and allowing vital independence.

unable to find suitable employment. Dogs are our levellers, allowing people of all ages and status to connect and make friends with other dog owners.

Research at the Davis School of Veterinary Medicine in California found that people with Alzheimer's suffered less

Dogs can be trained when still puppies. Once trained, they provide invaluable help in everyday life.

and partially-sighted people across the UK through the provision of guide dogs, mobility, and other rehabilitation services (see http://www.guidedogs.org.uk/).

Hearing Dogs for Deaf People is an organisation that trains dogs to be our ears, and was launched in the UK in 1982. The organisation trains dogs to alert deaf people to important sounds and danger signals in the home, work place and public buildings. It has its own breeding programme, and those breeds of dog which in particlar make good guides are Labradors, Golden Retrievers, Cocker Spaniels, Miniature Poodles, and Cavalier King Charles Spaniels.

Dogs are trained to respond to a variety of different sounds such as an alarm clock, doorbell, telephone and smoke alarm. With their noses they touch their owner to get their attention, then nudge to direct them to the cause of the sound. The dogs not only afford their owners invaluable help and protection, they also provide independence, confidence, and loving companionship. Being deaf can be very isolating, so a constant, loyal and attentive canine companion can be very therapeutic (see http://www. hearingdogs.org.uk/).

Pets as therapy (PAT dogs)

This organisation, founded in the UK in 1983, is a community-based charity providing therapeutic visits by volunteers and their dogs and cats to hospitals, hospices, nursing and care homes, special needs schools, and other establishments.

There are on average 4500 PAT dogs who visit the bedsides of around 130,000 people. The therapeutic value of having an animal to cuddle and talk to has been shown to help in times of distress and illness, as well as those with special needs, and inmates of residential institutions.

An American study carried out in residential homes for senior people had a human visitor, a human visitor and a dog, and just a dog visit three distinct groups of residents, after which the groups completed questionnaires to identify which type of visit they preferred. Not surprisingly, the most popular vistor was the dog on his own ...

Why *do* we love our dogs so much? The next chapter has the reasons ...

stress and had fewer anxious outbursts when there was a pet at home. Dogs can also have a calming effect on care-givers.

Dogs helping people

The value of the very many ways in which dogs help us has been recognised for many years. The Guide Dogs for the Blind Association (known as Guide Dogs) is a British charitable organisation that was founded in 1934, which provides independence and freedom to thousands of blind

I once had a client in my psychotherapy practice who, when asked what her goal in life was, answered: "I want to be as happy as my dog." I thought this was a pretty good goal, and asked her what it was about her dog's life that she found so appealing.

She told me that her dog –

- shows real enthusiasm for life
- doesn't waste time questioning what he does
- is entirely transparent: what you see is what you get
- gives unconditional love: "His affection and love for me is not dependent on my achieving things, behaving in certain ways, dressing 'appropriately', saying the right things, or promising to change"
- is non-judgemental: "He doesn't judge me, just takes me as I am. He doesn't berate me for the things I haven't done; the mistakes I have made or the decisions I have reached"
- doesn't criticise
- doesn't make fun of others
- is never sarcastic
- doesn't question himself
- is happy in his own fur
- doesn't lie
- is spontaneous
- lives for today and in the moment
- accepts what life has to offer
- shows his emotions openly
- doesn't compromise
- makes his needs clear

Many people would love to be as happy as their dog!

- has absolutely no hidden agenda
- lives a balanced life
- embraces his inner child
- doesn't suffer fools
- is always loyal
- is protective
- tunes in to others – knows if they are sick, anxious, unhappy, angry; recognises children as small people; picks up on those who are wary or nervous around dogs
- shows physical affection – wants to be close
- likes one-to-one human contact
- doesn't take offence or bear grudges
- solves problems and achieves goals – can be tenacious and focused
- is philosophical and stoical
- has a Buddhist philosophy – shows enlightened self-interest
- is confident when boundaries are clear and his living situation is secure
- doesn't waste time worrying
- doesn't dwell on issues
- does not suffer guilt
- is interested in new alliances with others of his species

The honest and unconditional love our dogs give is a fundamental reason why we value their companionship so much.

- is adventurous
- doesn't try to change who he is – he knows his colours suit him

All of the reasons she gave were a terrific testimony to her dog – and *the* dog – capturing his spirit, and going some way toward explaining why we love dogs as company. From the baby sleeping peacefully alongside the family dog; small children sharing their play with dogs; a young boy or girl learning responsibility as they care for their first dog; a comforting confidant in the teenage years, when the rest of the world seems at odds; through to a close and loving partner later in life – a dog is the very best companion.

Combining the life you share with your dog with another person, such as a partner, brings its own challenges, but wanting to be together, and the tolerance,

Our dog is happy in his own skin; happy simply to be himself, which a lot of us would like to be!

A dog can provide companionship and love throughout our lives, offering different kinds of support every step of the way.

Roxy came into a new place filled with lights, strangers, and paper flooring, but was completely at ease; happy to interact freely with her owner.

understanding and love that will facilitate this, can be the process that binds you all and cements your relationship.

"Love me, love my dog" – as true today as it ever was.

Harmonious living for your dog and you

Sometimes we concentrate so hard on training our dog we forget that it is ourselves who have the biggest influence on how he or she behaves. And the same is true of the various situations we find ourselves in.

A lovely lady called Suzanne brought her beautiful Husky, Roxy, by ferry to Falmouth and my house to take part in a photo session for this book.

Roxy had been fabulous: playing, sitting and generally being a great model throughout the photo shoot, even though she had never been to my house; had never met me or Tom, the photographer. She seemed completely happy and relaxed.

At one point, the noise her paws made on the set paper worried her, so we threw down a soft blanket to eliminate the noise and feel of the paper underneath and she was fine.

Suzanne told me that, the previous day, a Dog Whisperer had visited at her request as she had been concerned about a couple of uncharacteristic displays of behaviour from Roxy, who was normally great on walks, hardly ever barked, and was apparently content and settled. Suzanne explained she had a new neighbour who she really got along with, and they were excited about their dogs first meeting each other. They hadn't yet put up fences between the properties, and the dogs met for the first time in Roxy's garden. The neighbour was a little concerned about the first encounter between the dogs, and, whilst they were chatting, an altercation broke out between Roxy and the neighbour's dog, accompanied by much barking. Suzanne was very upset about this, and apologised. Later that week, whilst walking Roxy, Roxy suddenly began barking at another dog, so Suzanne decided that help was called for.

The Dog Whisperer initially did some preparation work with Roxy, and then asked that the neighbour and her dog come to the house. When the two dogs met they were curious but calm and friendly, and Roxy has been fine since.

Suzanne and I talked about how maybe Roxy had picked up on the neighbour's anxiety about the dogs meeting on what was normally Roxy's exclusive territory. This anxiety had possibly then transferred to Suzanne, which Roxy would have picked up on. I guess Roxy was doing the natural and instinctive thing in insecure situations: barking to warn and protect her owner. The later incidence of barking could have been due to Roxy detecting Suzanne's underlying fear that Roxy would bark again – which Roxy did as she again felt the situation was threatening: a self-fulfilling prophecy.

I think that many of us have experienced a similar situation with our dogs when, perhaps, we meet a friend or have someone to the house who is nervous around dogs, which, in turn, unsettles and agitates the dog to the extent that they may bark and act out of character; confirming the nervous individual's feeling that dogs are something to be afraid of! A situation like this can be embarrassing, causing us to become agitated and anxious, and perhaps feeling that we should be able to prevent our dog from barking. But this really becomes a vicious circle of your dog then reacting to *your* tension, further unsettling him or her, which makes *you* more stressed, etc, etc! In reality, your dog will probably settle as soon as the visitor has gone – and so can you.

Our dogs' ability to sense unease within us or a situation can result in a change in behaviour. We need to understand the instinctive impetus behind this to enable our training methods to be successful (and right).

A dog's-eye view

What are the important things to consider when looking at the world our dogs live in and share with us through *their* eyes. The basics are, our dogs need to know that we are in control and will take care of them, and that they can rely on us for food, shelter, and survival.

This is just the same, of course, with human babies, who are dependent on their care-givers for at least sixteen years from birth: the only animals who remain dependent for this period of time. And just like children, if they feel that

their survival and position are threatened, our dogs can feel insecure and anxious.

What are some of the things our dogs need to make them feel happy, safe and secure – and how do these compare with the things that we, as people, need to achieve the same effect?

Us	Our dog
Food, water, shelter, rest	Food, water, shelter, rest
Safety and security	Safety and security
Friends, family, significant relationships	Stability and consistency in relationships; trust
Sense of belonging	A secure place in the human-canine social group
Creativity, fun and playtimes	Playtimes, fun and relaxation

"My goal in life is to be the person my dog thinks I am" is a quote that American Tara Hedman has on a plaque on her wall. Tara loves her little ShihTzu, Murphy, and shared what he means to her in an article that she kindly gave me permission to quote. Here are some of the things she says about her little dog –

He thinks I am worth protecting
Tara describes how Murphy (who, she says, is a lover, not a fighter) chased after another dog in the street who had nipped her, catching up with him and pinning him against a porch to keep him away from her.

He thinks I am worth comforting
"Whenever I get teary or upset, he gently crawls onto my lap and curls up."

He thinks I am worth loving
"Relaxing in my hammock, Murphy asks to be lifted up, settling beside me and fast asleep within seconds."

He thinks I am worth missing
"When Murphy sees my suitcases, I hear his little doggie heart saying 'Don't go!'"

and others – and that includes our beloved four-legged friend, the dog.

I would like to close with a poem, written by the partner of a friend who runs a writing weekend retreat from a beautiful collection of stone cottages in deepest Devon. John was not taking part in the writing course: he was an engineer by profession, and not really 'into' writing. He was very happy, however, to take us on a walk around his beloved valley, showing us the river, its nature, and the animals they kept. John's daughter, Jade, who had been cooking for us, joined us with her friendly black Labrador, Bess. After a couple of hours' walking, we took time out to

Our dogs rely on us to provide support, affection, and direction, just as our children do. They crave the security that a balanced and well-adjusted 'leader' can provide.

Opposite: The simple, everyday things we sometimes take for granted can make our dog so happy, and his zest for life and joyous spirit can help us in so many ways.

He thinks I am worth being close to

"When I do the dishes he sits on my feet; when I'm in bed he snuggles close to my head."

We all have occasions in our lives when we feel unloved, lost and alone; when it seems that not a soul cares about us. But even at the darkest of times – when we don't even believe it ourselves – our dogs let us know that we are valued, loved – and most definitely not alone. Our dogs not only share our lives, they very often know more about us than anyone else; our comings and goings; hopes and dreams, highs and lows. And they accept all of what we are unconditionally.

Learning more about ourselves and what makes us tick will enable us to live more harmoniously with ourselves

Our dogs have the ability to make us feel happier about ourselves: they think such a lot of us; so should we.

Your dog and you – understanding the canine psyche

enjoy our surroundings in the warm autumn sunshine, and do some writing if we felt so inclined.

Getting together later that evening to discuss the day's writing around the campfire John had built, he quietly admitted he had written a poem, the first in his life. It was about Bess, he said; written from her perspective. I asked if I could see it, and the next day, John wordlessly handed me a piece of paper containing the poem.

I felt that John's words summed up what it was I wanted to say in this book, and I asked if I might include it –

Bess in the valley
I joined with the writers' group in their walk round the valley
They didn't bring a ball

They stopped in silence to listen
I joined in – I heard the birds sing in the sunshine

Then a distant shotgun
Shotguns are loud
Shotguns are dangerous
Shotguns can kill a small dog in a valley
I didn't like listening

Why didn't they bring a ball?
They throw a ball

I fetch a ball
They say "Good dog!"
They throw a ball
I fetch a ball
They say "Good dog!"
They throw a ball
I fetch a ball
They say "Good dog!"

Life is simpler with a ball

John Bartlett
Watermill Cottages, South Devon
Writers' Group visit
www.watermillcottages.co.uk

Asking only to be a part of our lives,
our canine companions give it all back,
a million times over!

The Dog Whisperer: the gentle way to train your best friend by the man who speaks dog • Graeme Sims • Headline (2009)
• ISBN: 9781929242368

Think Dog • John Fisher • Octopus Books (2012) • ISBN: 9781844037094

The Complete Dog Massage Manual: Gentle Dog care • Julia Robertson • Hubble & Hattie (2010) • ISBN: 978184584322-9

The Genius of Dogs • Brian Hare & Vanessa Woods • Oneworld (2013) • ISBN: 9781851689859

How to Speak Dog • Stanley Coren • Pocket Books (2005) • ISBN 1416502262

Do You Look Like Your Dog? • Gini Graham Scott • Changemakers (2011) • ISBN 97811466292161

Clever Dog. Understand What Your Dog is Telling You • Sarah Whitehead • Harper Collins (2013) • ISBN 9780007488544

The Secret language of Dogs: how to communicate effectively with your dog • Heather Dunphy • Apple Press (2011)
• ISBN: 9780956763907

Life Skills for Puppies – Laying the foundation for a loving, lasting relationship • Helen Zurich & Daniel Mills • Hubble & Hattie
(2012) • ISBN: 9781845844462

Dog Language • R Abrantes • Dogwise • ISBN: 9780966048407

Dog Speak: recognising and understanding behaviour • Christiane Blenski • Hubble & Hattie (2012) • ISBN: 9781845843847

Clever Dog! Life skills from the world's most successful animal • Ryan O'Meara • Hubble & Hattie (2011) • ISBN: 9781845843458

Know Your Dog – The Guide to a Beautiful Relationship • Immanuel Birmelin • Hubble & Hattie (2010) • ISBN: 9781845840723

The Truth about Wolves and Dogs: Dispelling the myths of dog training • Toni Shelbourne • Hubble & Hattie (2012)
• ISBN: 9781845844271

Introduction to CBT at Work • Gill Garratt • Icon (2012) • ISBN: 9781848314191

Dillon

Jagger

Barney

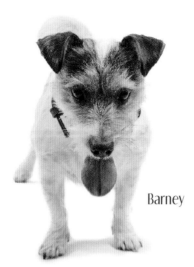

Roxy

Dudley

Ozzie

Lola

Ruby

Jack

Roxy

Sky

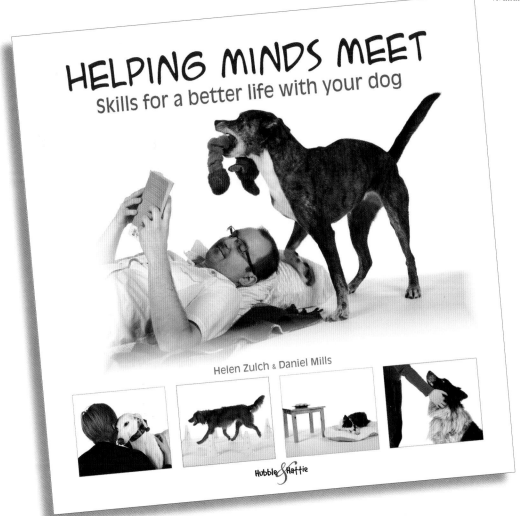

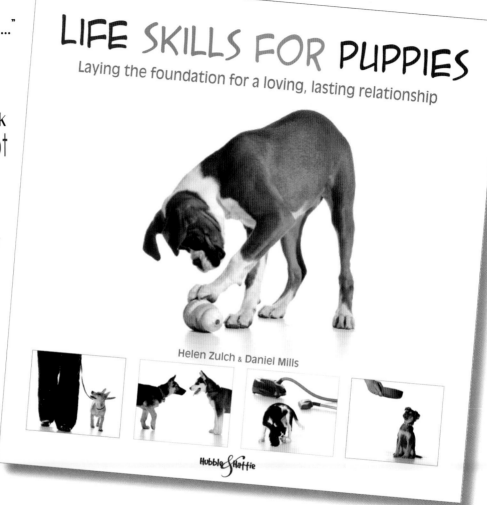

LIFE SKILLS FOR PUPPIES
Laying the foundation for a loving, lasting relationship

Helen Zulch & Daniel Mills

Hubble & Hattie

Puppy education from the puppy's perspective! Presenting the key skills that a dog needs to cope with life, this ground-breaking book, written by professionals in the field, aims to assist owners develop a fulfilling relationship with their puppy, helping him to behave in an appropriate manner and develop resilience, whilst maintaining good welfare. The skills taught are incorporated into everyday life so that training time is reduced, and practising good manners and appropriate behaviour become a way of life.

205x205mm • 96 pages • 121 colour images • paperback plus flaps • ISBN 9781845844462 • £12.99*

Index